Agroecology: Sustainable Ecosystem

Anand M. R.

Poojitha Kommireddy

Seenappa C.

Kalyana Murthy K. N.

Ningaraju G. K.

CRC Press is an imprint of the
Taylor & Francis Group, an **informa** business

Elite Publishing House

First published 2024
by CRC Press
4 Park Square, Milton Park, Abingdon, Oxon, OX14 4RN

and by CRC Press
2385 NW Executive Center Drive, Suite 320, Boca Raton FL 33431

CRC Press is an imprint of Informa UK Limited

British Library Cataloguing-in-Publication Data
A catalogue record for this book is available from the British Library

Print edition not for sale in India.

ISBN: 9781032627755 (hbk)
ISBN: 9781032627762 (pbk)
ISBN: 9781032627779 (ebk)

DOI: 10.4324/9781032627779

Typeset in Adobe Caslon Pro
by Elite Publishing House, Delhi

Contents

Preface

At global level, the planet earth grappled with serious environmental and social changes, it became obvious that preserving the mother earth is the most critical topic of our times. Today, it is extremely important to have a understanding of ecological issues. This is because the human economy is engaged in a wide range of activities that are causing enormous damage to the ecosystems that sustain both our species and Earth's legacy of biodiversity. All around us, this is witnessed by pollution, climate warming, land sliding, deforestation, the degradation of agricultural soil, extinctions and endangerment of species, and other damages. Systematic understanding and execution of agro ecology address all these issues. Agro- ecology literacy can be defined as: "the degree to which people have an objective and well-informed understanding of crop production with environmental issues".

In the Indian context, Green revolution was a milestone in achieving food security of independent India. The increased food grain production was achieved mainly through the use of high yielding varieties and external inputs. These high yielding varieties are highly responsive to fertilizers and to some extent amenable to pests and diseases. This in turn demanded increased use of fertilizers and plant protection chemicals. The unscientific and indiscriminate use of agro-chemicals has affected the soil health and degraded agro ecosystem adversely and brought down the immunity of soil and productivity of crops. In addition, it has resulted in the over exploitation of soil leading to nutrient imbalance. Moreover, a negative impact of chemical agriculture on environment and human health has been reported and documented. Increased environmental awareness and health consciousness promoted the scientists and planners to think about sustainable farming practices as an alternate way for healthy agriculture with a protection to environment and human health. This issue can be very well addressed through eco friendly methods or ecosystem services.

We sincerely hope that this book will be very much useful to the students and academicians. We also welcome all suggestions for further refinement of this publication. We whole-heartedly acknowledge the inputs given by the contributors to various sections.

Authors

About the Authors

Dr. Anand M.R. presently working as Associate professor at University of Agricultural Sciences, GKVK, Bangalore. He did his UG, PG and Ph.D. from UAS, Bangalore. He is serving in the university from last 20 years in teaching, research and extension wings. He was recipient of Young Scientist award from CHRIST University, Bengaluru and many other professional awards from various organizations.

Dr. Kommireddy Poojitha is currently working as Teaching Associate at S.V. Agricultural College, Tirupati. She did her UG from S.V. Agricultural College, ANGRAU, Tirupati and PG and Ph.D from UAS, GKVK, Bangalore. She also did PG Diploma in Agricultural Extension Management from MANAGE. She was recipient of university gold medal and Doctoral Research Fellowship from Department of Science and Technology, Govt. of Karnataka.

Dr. Seenappa C. presently working as Professor at University of Agricultural Sciences, GKVK, Bangalore. He did his UG, PG and Ph.D from UAS, Bangalore. He is serving in the university from last 16 years in teaching at different levels. He was recipient of Dr. B.R.Ambedkar University gold medal at Doctoral level.

Dr. Kalyana Murthy K.N. presently working as Professor and University Head of Department of Agronomy at University of Agricultural Sciences, GKVK, Bangalore. He did his UG, PG and Ph.D from UAS, Bangalore. He is serving in the university from last 30 years in teaching, research and extension wings. He was recipient of ICAR best teacher award.

Dr. Ningaraju, G.K. obtained his graduation from College of Agriculture, Hassan, Post graduation in Agronomy from Kerala Agricultural University and Ph.D. from the University of Agricultural Sciences, GKVK, Bangalore, Karnataka. With the Best KAU Thesis Award, he earned his master's degree, 13 research papers, several brief research notes and extension bulletins have been published by him. He is currently working as an Assistant Professor of Agronomy (C) at the College of Agriculture, UAS, GKVK, and Bangalore.

Chapter - 1

Concept of Crop Ecology

The Science of ecology is relatively a young branch of biology which deals with interacting system of organisms and their environment. The term Ecology (old spelling *Oekologie*) has been derived from two '**Greek**' words, *Oikos* meaning house or dwelling place and *Logos* meaning the study. Therefore, ecology is the study of organisms in their natural home or habitat. So, Ecology is the study of organism 'at home'. Ecology is generally defined as the study of plants and animals in reciprocal relationship with their environment or external world. It is one of the basic divisions of biology like others-morphology, physiology, cytology, genetics, taxonomy etc., but it differs from them essentially in two ways:

1. It always comprehends along with organisms and the non-living environment.
2. It deals with system of levels higher than the organisms in the biological spectrum. (*i.e.* Organization of systems of genes, cells, organs, organ systems, organisms, population and communities).

Ecology concerns mainly with the study of populations, communities, ecosystem and biosphere. The biosphere is not a homogeneous structure and is composed of numerous ecosystems/habitats which differ in their composition and functional attributes. By system, it means a unified whole made of regularly interacting and interdependent components. Organisms on the one hand and non-living parts like material and energy *etc.* on the other hand form the major components of the system which regularly interact among themselves and remain interdependent.

Population systems formed by assemblage of a large number of individuals of any one species interact with their environment and the ecosystems formed by interaction of assemblage of a variety of populations *i.e.*, communities with their environment are the main structural and functional entities, studied in ecology. Naturally, populations and communities of all kinds of taxonomic entities like algae, fungi, angiosperms, arthropoda, mammalian are studied with respect to their functions in the ecosystem. The largest ecological system which includes all the organisms of earth and the total environment is called biosphere or ecosphere.

Environment constitutes conditions influencing development or growth and quality of life. It is anything and everything which influences the life activities of the organisms. Thus, any habitat encompassing space possesses plants and animals which constitute the biotic environment while the abiotic environment is constituted by air, water, mineral from soil, and solar energy. The biotic and abiotic environment interacts with each other and form dynamic systems in the habitat

Ecology is not synonymous with environmentalism, natural history, or environmental science. It overlaps closely with the related sciences of evolutionary biology, genetics, and ethology. An important focus for ecologists is to improve the understanding of how biodiversity affects ecological function. Ecologists seek to explain the life processes, interactions, and adaptations, the movement of materials and energy through living communities, the successional development of ecosystems and the abundance and distribution of organisms and biodiversity in the context of the environment.

In 1859, **G.S. Hilaire** used the term *ethology* which refers to the 'study of relationship between organisms and environment'. **Reiter** (1868) introduced the term *Oekologie* in literature and **E. Hacckel** (1969) put the first precise definition of ecology. Later, the community aspect in ecology was introduced. Notable contributions to plant and animal communities were made by S.A. Forbes (1887), Warning (1907) and Clements (1916).

The traditional definition of **ecology** is 'the study of an organism and its environment'; however, different ecologists have defined it variously. **Ernst Haeckel** (1869) defined ecology as, 'the total relation of an animal to both its organic and its inorganic environment'. In 1936, **Taylor** defined ecology as 'the science of all the relations of all organisms to all their environments'. **Charles Elton** (1947) in his pioneer book 'Animal Ecology' defined ecology as 'scientific natural history.' Although this definition does point out the origin of many of our ecological problems, yet it is much broad and vague like Haeckelian definition of ecology. **Allee** *et al.*, (1949), in their definition of ecology, clearly emphasize the all-encompassing character of this field of study. According to them **ecology** may be defined broadly as 'the science of the interrelation between living organisms and their environments, including both the physical and biotic environments, and emphasizing interspecies as well as intra-species relations'. Though, **F.J. Vernberg and W.B. Vernberg** (1970) completely agree with **Allee** *et al.* definition, yet there are certain ecologists which are not satisfied with this definition and have provided their own definitions of ecology. For instance, **Andrewartha** (1961) defined ecology as 'the scientific study of the distribution and abundance of organisms'. **G.A. Petrides** (1968) has defined ecology as 'the study of environmental interactions which control the welfare of living things, regulating their distribution, abundance, production and evolution'. **Eugene Odum** (1963, 1969 and 1971) has defined ecology as 'the study of the structure and function of ecosystems'.

Apart from all these definitions, there were few more definitions of ecology. They are

- The branch of biology concerned with the relations between organisms and their environment.
- Study of the interaction between living things and their physical, chemical and biological environment.
- The branch of science studying the interactions among living things and their environment.
- Science of the relationships between plants, animals and their environments.
- The study of the interactions of organisms with their physical environment and with one another.
- The study of the interactions of living organisms with each other and with their environment.

Ecology deals with

- The temporal changes in the occurrence, abundance and activities of organisms
- The interrelations between organisms, communities and populations
- The structural adaptation and functional adjustments of organisms to the change in environment

» The behavior of organisms under natural environment and the productivity of these organisms

» Energy and other natural resources to mankind and

» The development of interactive models for analytical or predictive purposes.

Chapter - 2

Historical Developments

In Indian writings such as Vedic and Epics, we may find references to ecological thoughts and **'Chakra'** nature worshiper described the importance of Vayu (gases and air), Jala (water), Desha (topography) and Kala (time) in regulation of plants life. The Greek philosophers and scientists like Hippocrates (father of medicine), Aristotle, Theophrastus and Reaumur described in their writings on natural history, the habits of animals and plants growing in different areas. C. **Linnaeus** (1707-1778) in his book **"Natural History"**, in 1756 made notable contributions to ecology.

The major milestones in the growth and development of the subject of ecology are:

1. **Alexander von Humboldt and Aime Bonpland** (1805) has written an essay on geography of plants.

2. In 1825, the French naturalist, **Adolphe Dureau de la Malle** used the term *societe* about an assemblage of plant individuals of different species.

3. **The notion of biocoenosis by Wallace and Mobius**

Alfred Russel Wallace, contemporary and competitor to Darwin, was first to propose a "geography" of animal species. Several authors recognized at the time that species were not independent of each other and grouped them into plant species, animal species and later into communities of living beings or biocoenosis. The first use of this term is attributed to Karl Mobius in 1877.

4. Warming and the foundation of ecology as a discipline

Eugen Warming devised a new discipline that took abiotic factors, that is drought, fire, salt, cold etc., as seriously as biotic factors in the assembly of biotic communities. Warming gave the first university course in ecological plant geography.

5. Darwinism and the science of ecology

The roots of scientific ecology may be traced back to Darwin. It comes from his work *On the Origin of Species* which is full of observations and proposed mechanisms that clearly fit within the boundaries of modern ecology. The term 'ecology' was coined in 1866 by 'Ernst Haeckel' a strong proponent of Darwinism.

6. Early 20th century ~ Expansion of ecological thought

By the 19th century, ecology blossomed due to new discoveries in chemistry by Lavoisier and De Saussure, notably the nitrogen cycle. After observing the fact that life developed only within strict limits of each compartment that makes up the atmosphere, hydrosphere, and lithosphere, the Austrian geologist 'Eduard Suess' proposed the term 'biosphere' in 1875. Suess proposed the name biosphere for the conditions promoting life, such as those found on earth, which includes flora, fauna, minerals, matter cycles, etc.

7. **In the 1920s Vladimir I. Vernadsky,** a Russian geologist, detailed the idea of the biosphere in his work "The biosphere", 1926. It was he who described the fundamental principles of the biogeochemical cycles. He thus redefined the biosphere as the sum of all ecosystems.

8. The Ecosystem: Arthur Tansley

Over the 19th century, botanical geography and zoogeography combined to form the basis of biogeography. This science, which deals with habitats of species, seeks to explain the reasons for the presence of certain species in a given location. It was in 1935 that **Arthur Tansley**, the British ecologist, coined the term **ecosystem**, the interactive system established between the biocoenosis (the group of living creatures), and their biotope, the environment in which they live. Ecology thus became the science of ecosystems.

9. Eugene Odum

Tansley's concept of ecosystem was adopted by the energetic and influential biology educator Eugene Odum. Along with his brother, Howard Odum, Eugene Odum wrote a textbook which educated more than one generation of biologists and ecologists all over the world.

Eugene Odum, published his popular 'Ecology textbook' in 1953. He became the champion of the ecosystem concept. This ecosystem science dominated the International Biological Program of the 1960s and 1970s, bringing both money and prestige to ecology.

10. Ecological Succession - Henry Chandler Cowles

At the turn of the 20th century, **Henry Chandler Cowles** was one of the founders of the emerging study of "**Dynamic ecology**", through his study of ecological succession.

11. Ecology's influence in the social sciences and humanities

Human ecology has been a topic of interest for researchers, after 1920. Humans greatly modify the environment through the development of the habitat (in particular urban planning and growth), by intensive exploitation activities such as logging and fishing, and as side effects of agriculture, mining, and industry. Besides ecology and biology, this discipline involved many other natural and social sciences, such as anthropology and ethnology, economics, demography, architecture and urban planning, medicine and psychology, and allied areas. The development of human ecology led to the increasing role of ecological science in the design and management of cities.

12. Ecology and global policy

Ecology became the central core of the world's politics by 1970. It was mainly due to role played by the UNESCO which launched a research program called *Man and Biosphere*, with the objective of increasing knowledge about the mutual relationship between humans and nature. A few years later it defined the concept of 'Biosphere Reserve'. In 1972, the United Nations held the first international Conference on the Human Environment in Stockholm, prepared by Rene Dubos and other experts. This conference was the origin of the phrase "Think Globally, Act Locally".

Why it is important to study ecology?

Existence in the world is made up of living and non living things. The two groups have to coexist in order to share the resources that are available within the environmental ecosystem. To understand about this mutual co relationship we need to study and understand ecology. Survival of all organisms is actualized due to ecological balance. Various species survive because favourable ecosystems were created around them. One core goal of ecology is to understand the distribution and abundance of living things in the physical environment. Attainment of this goal requires the integration of scientific disciplines inside and outside of biology, such as biochemistry, physiology, evolution, biodiversity, molecular biology, geology, and climatology. Some ecological research also applies many aspects of biology, geology, chemistry and physics, and it frequently uses mathematical models. Ecologists study these relationships among organisms and habitats of many different sizes, ranging from the study of microscopic bacteria growing in a fish tank, to the complex interactions between the thousands of plant, animal, and other communities found in a desert. Ecologists also study many kinds of environments. For example, ecologists may study microbes living in the soil under your feet or animals and plants in a rain forest or the ocean.

Chapter - 3

Levels in Ecological Organization

Ecologists study nature on different levels, from a local to a global scale. These levels reveal the complex relationships found in nature (**Fig.1**).

i. **Organism:** An organism is an individual living thing and is the basic unit of study (Ex: An alligator). It includes the study of the form, physiology, behavior, distribution and adaptation of organism in relation to environment.

ii. **Population:** A group of organisms consisting of a particular species that live in defined area and interact with each other is called population (Ex: all the alligators that live in a swamp). It includes the study of interaction between populations and intra-specific relationships. A population is a group of organisms of the same species that lives in one area as such.

iii. **Community:** A community is a group of different species that live in an area and interact with each other (Ex: groups of alligators, turtles, birds, fish and plants that live together in a swamp). It includes the study of structure and composition of community and inter-specific interactions between members of community.

iv. **Ecosystem:** An ecosystem includes all the communities of the organisms as well as the climate, soil, water, rocks, and other non-living things in a given area. Ecosystems can vary in their size (Ex: communities of alligators, turtles, birds, fish and plants that live together in a swamp along with abiotic components constitute fresh water ecosystem). Here, communities in relation to the structure of its ecosystem-nutrients cycling, climate, energy flow etc… are studied.

v. **Biome:** A biome is a major regional or global community of organisms. A biome is a large community of plants and animals that occupies a distinct region (may be grassland, tundra, aquatic, desert and forest). Biomes are usually characterized by the climate conditions and plant communities that thrive there.

vi. **Biosphere:** It is the part of the planet where life exists and a biosphere is a collection of all the biomes.

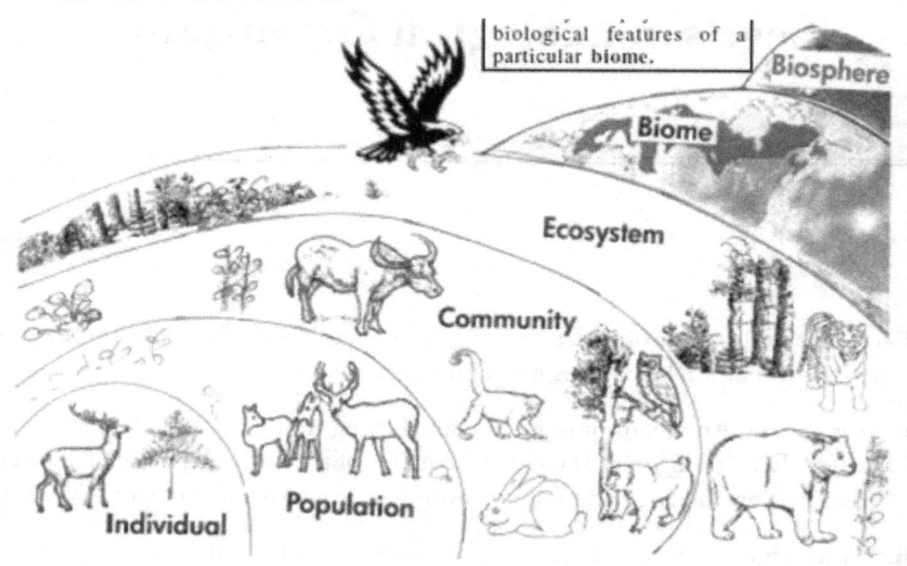

Fig. 1: Levels of ecological organization

Levels of study of Ecology

I. Organism level

A. Plant ecology

1. Autecology - Dealing with the individual organism or species in relation to its environment.

2. Population ecology – Study of population

3. Synecology - focused at understanding the interactions of groups of organisms or species within a community

B. Animal ecology

II. Habitat or ecosystem level

A. Terrestrial

(1) Forest ecology

(2) Grassland ecology

(3) Desert ecology

(4) Marsh or wetland ecology

B. Aquatic

(1) Marine ecology

(2) Fresh water ecology

III. Applied aspects of ecological studies

(1) Conservation ecology

(2) Pollution ecology

(3) Crop ecology or Agricultural ecology

3.1. Divisions of Plant Ecology

Plant ecology is further divided into Autecology and Synecology

Autecology

This is also known as ecology of individuals, where we study the relation of individual species to its environment. Thus at a given time, emphasis is given on the requirements and reaction of an individual species together with the influence of environment upon it. With an autecological approach, individual species are the units of study. These are studied for details of their geographic distribution, morphology, taxonomic position and life-cycle etc., along with the various ecological factors which might influence different stages of their life cycles.

Synecology

Under natural conditions, however, organisms – plants, animals, microbes etc., live together as a natural group affecting each other's life in several ways. Thus, more complex situations exist where the units of study, instead of single organisms are groups of organisms known as a community. Such an approach where units of study are groups of organisms is called synecological approach. Schroeter and Kirchner (1896) introduced the term synecology.

Depending upon the conditions as these exist, synecology may deal with-

1. Population ecology: It is recently developed field, where the units of study are pure stands of individuals of a single species - population, As a result of aggregation of these individuals, it becomes desirable to study the interdependencies between them, and the populations are studied in terms of their size, growth rates etc., which are chiefly governed by the interactions of the members of population (**Fig. 2**). Thus, population ecology is the study of such and other similar relationships of group of organisms. The main job of population ecologist is, "Why is this population of a particular density?" To answer this and other questions he studies competition, usually between population from the same trophic level (herbivores competing for same grass). Population ecology is also concerned with communities. A population ecologist also studies interactions between populations of different species in a community. For instance, study of prey-predator interactions between members of adjacent trophic levels of a community.

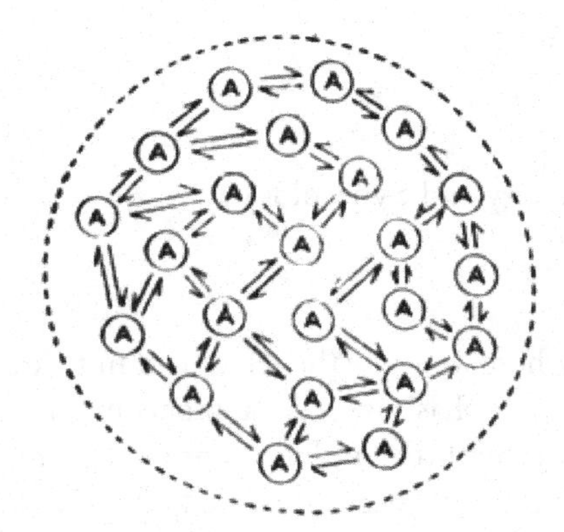

Population of species A

Fig. 2: Diagrammatic sketch showing the population of species 'A' where its individuals interact with each other - **Population ecology**

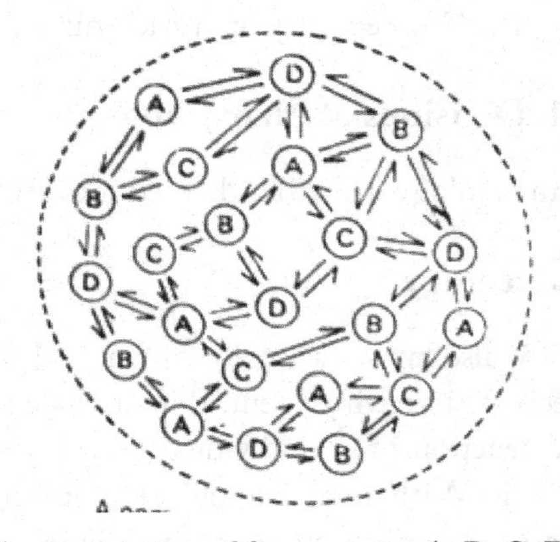

A community of four species A, B, C, D

Fig. 3: Diagrammatic sketch showing a community of four different species A, B, C & D interacting with each other - **Community ecology**

2. Community ecology: In contrast with population ecology, here the units of study are groups of individuals belonging to different species of plants as well as animals. The living (biotic) components of the community are studied mainly for the nature of interdependencies between individuals of different species (**Fig. 3**). Major concerns of community ecologist are, "Why is this community of a particular diversity? Why does a particular community occur at a given location? How communities interact and how these change through time?"

3. Biome ecology: In nature, we generally find that there may exist a complex of more than one community, some in their climax stages and others in different stages of succession, and these all communities grow under more or less similar climatic conditions in an area. Thus in biomes, interactions between different communities of area as are studied (**Fig. 4**).

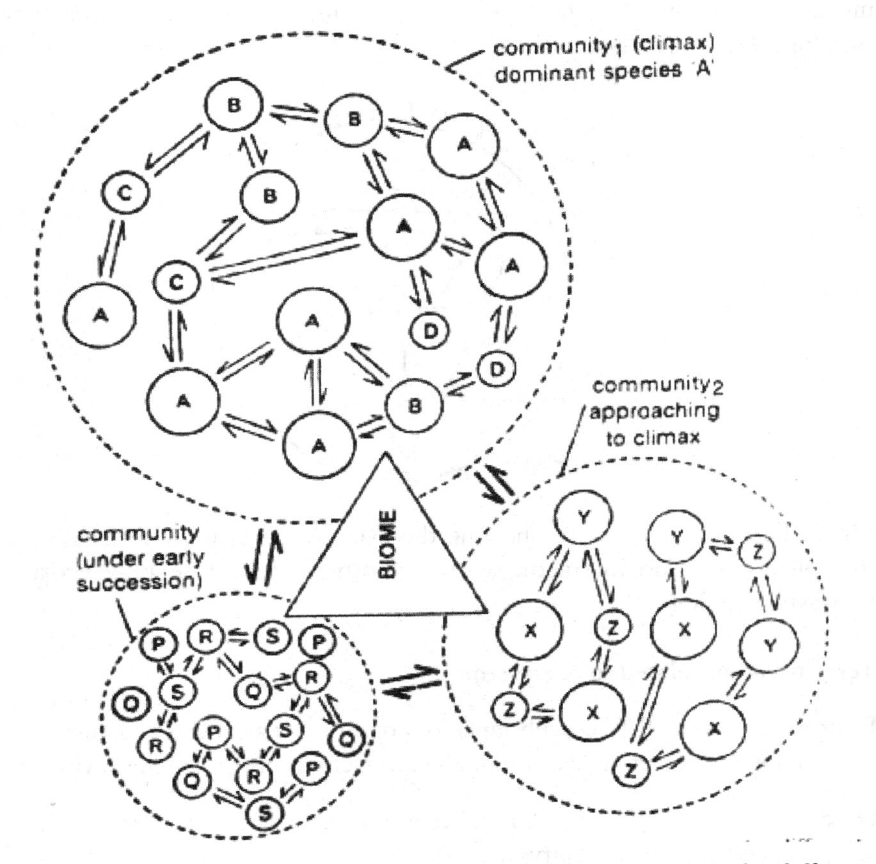

Fig. 4: Diagrammatic sketch showing a biome of three communities under different stages of development and interacting with each other - **Biome ecology**

4. Ecosystem ecology: This has been the most recent development in ecology. It is established that not only living (biotic) but also non-living (abiotic) components of the nature interact with each other. These interacting biotic and abiotic components then interact with each other to form an integrated system-ecosystem or ecological complex or ecological system (**Fig. 5**). Thus, it becomes the most complicated synecological approach to the ecology of an area, where the units are the whole system-living as well as non-living components. Here, we mainly emphasize the similarities and differences in food relationships among living organisms and various forms of energy supporting their life. This has as bio-energetic approach in modern ecology.

Ecosystem ecology thus emphasizes the movements of energy and nutrients among the biotic and abiotic components of ecosystems. A major concern is, "How much and what rates are energy and nutrients being stored and transferred between components of an ecosystem?"

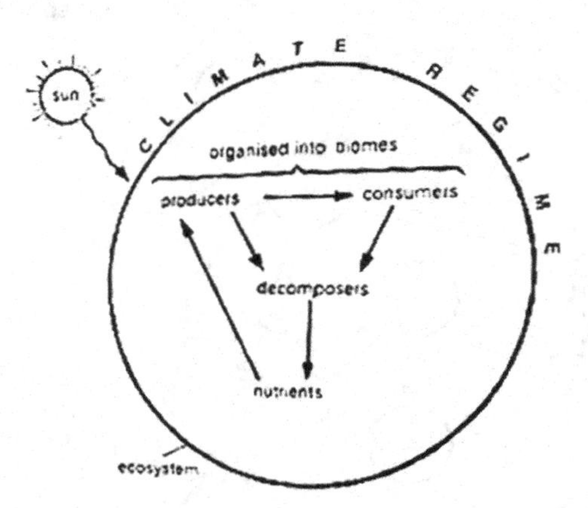

Fig. 5: Diagrammatic sketch showing the living (biotic) and non-living (abiotic) components of nature interacting with each other to form integrated ecosystem-**Ecosystem ecology**

Terminologies related to ecosystem:

Factor: Any external force, substance or condition that affects organisms in any way is known as a factor. The sum of all such factors constitutes the environment.

Diversity: Combination of the number of species and the number of individuals of each species in a community

Habitat: Habitat is the physical environment in which an organism lives. Each organism has particular requirements for its survival and lives where the environment provides for those needs. A habitat may support many different species having similar requirements.

Niche: In nature, many species occupy the same habitat but they perform different functions. The functional characteristics of a species in its habitat are referred to as "**niche**" in that common habitat. Habitat of a species is like its 'address' (*i.e.*, where it lives) whereas niche can be thought of as its "profession" (*i.e.*, activities and responses specific to the species). The term niche means the sum of all the activities and relationships of a species by which it uses the resources in its habitat for its survival and reproduction.

A niche is unique for a species while many species share the habitat. No two species in a habitat can have the same niche. This is because if two species occupy the same niche they will compete with one another until one is displaced. For example, a large number of different species of insects may be pests of the same plant but they can co-exist as they feed on different parts of the same plant.

Another example is the vegetation of the forest. The forest can support a large number of plant species as they occupy different niches: the tall trees, the short trees, shrubs, bushes and grasses are all part of the forest but because of varying heights they differ in their requirements for sunlight and nutrients and so can survive together.

Plant ecology: Plant ecology deals with plants in relation to their environments. The plant ecologist is concerned mainly with the habitats of plants and associations of plants or with the physiology of the plant or group of plants in a particular environment.

Crop ecology: Crop ecology may be defined as the ecology of crop plants. It is the study of the inter relationships between crop plants and the environment.

Ecological crop geography: Ecological crop geography deals with the broad distribution of crop plants and the underlying reasons for such distributions. The ecological crop geographer is concerned with more than the direct relationships of crop plants to their physiological environment. He must consider the points taken into account by the crop ecologist and in addition must recognize the operation of economic, political, historical, technological and social forces. These additional forces are grouped under the term social environment.

Thus, ecological crop geography maybe defined as the study of crop plants in relation to their physiological and social environments.

Chapter - 4

Crop Ecology

Agronomy is the art and underlying science of handling the crop plants and the soil substrate as to produce the highest possible quantity and quality of the desired crop product from each unit of land, soil, water and light with a minimum of immediate or future expense in labour and soil fertility. In standard dictionaries, agronomy is generally defined as "the management of land" and as "rural economy". The general public has learned that the term implies to the study of problems connected with the production of farm crops.

Two facts are in evidence from the attempts of defining agronomy: (a) the physiological and (b) the economic relationships. The present divisions of agronomic studies are in themselves indicative of the far reaching activities in this general field of agricultural research. The main lines are generally drawn along crops and soil studies. These divisions are subdivided into special phases even though the lines between crops and soil studies may not always be definite. Plants grow in the soil and results of soil treatment are generally measured by the plant responses.

Agricultural developments especially in recent years have brought out forcefully the necessity for what may be termed a world outlook on agricultural production. Agricultural production or any other form of production is influenced not only by local factors but to a great extent by world conditions too. The development of such a conception of agricultural production demands a broad outlook, it cannot confine itself to the physiological and mechanical phases of production in any one locality but must consider also the world economic and social forces influencing production of specified crop plants.

It is essential for the agronomist, in order to obtain a well rounded concept of his field, not only to consider local factors of production but also to become acquainted with the main factors determining the location of the centers of crop production within the confines of his own country and with the forces determining world centers of production.

Investigations during the past half century have set ecology and ecological relationships more and more on a scientific and experimental basis. To explain how organisms adapt themselves to a precise environment calls for a mustering of all available knowledge of plant morphology, anatomy and physiology. It is not too inclusive to say that most agronomic investigations touch very directly on the ecological relationships of crop plants. Soil investigations, work in crop breeding, variety testing, choice of special crops to meet certain conditions and numerous other agronomic projects are definitely based on ecology and ecological relationships.

Crop ecology may be defined as the ecology of crop plants. It is the study of the inter-relationships between crop plants and the environment. Factors controlling the habitat of crop during its entire life cycle such as climate, physiography, soil and biota are considered. The study of crop ecology is related to the study of agro ecosystems and the field of agro-ecology is not associated with any one particular method of farming, whether it be sustainable, intensive or extensive. Gliessman (1992) defined 'agro-ecology' as the application of ecological concepts and principles to the design and management of sustainable agricultural systems. The agro-ecological environment of a crop, land use or a farming system covers physical, chemical and biological aspects. There is need for integrating the overlapping ecological and environmental traits with sociological, economic, political and other cultural components of agriculture. In order to avoid confusion between the tasks of crop ecology and ecological crop geography, the study of the former should be confined to investigation of the relationships of crop plants to their physiological environments to the exclusion of the economic factors encountered in the production and distribution of a crop or group of crops. The effects of both physiological and economic factors on production and distribution of crops are treated in the field of ecological crop geography.

Cultivation of crops has become a part and parcel of our life. For getting best results in farms and fields ecological principles have to be applied. Crop ecology is therefore, become an important subject of study for successful agriculture. Crop ecology is concerned with the management of soil for growing of crops.

The best example of applied ecology is agriculture since the entire problem of crop production is an ecological problem. A particular crop ecosystem thrives only in particular climate and edaphic conditions. The first job of a crop ecologist is therefore to select a plot of land in a place where the climatic and soil conditions are suited to the particular crop. Soil fertility is getting depleted due to continuous cropping. A crop ecologist applies his knowledge of the nutrients requirements of the crops in such situation. By applying manures and fertilizers he restores the fertility of the soil and once again a good yield can be expected. A crop ecologist should also pay attention to the improvement of soil condition by adopting methods for soil conservation and for control of soil erosion. Maintenance of soil fertility can be ensured by the application of the ecological principle of crop rotation. The nutrient cycling of the ecosystem is thus balanced.

The crop ecologist should also know the ecology of weeds, pests and diseases to check their adverse effects on the crop plants. An agronomist has therefore to be well versed in plant ecology, for getting highest yields from his crops. Successful soil management coupled with proper land use enables a crop ecologist to get the best of an agricultural crop.

Crop ecologists do not unanimously oppose technology or inputs in agriculture but instead assess how, when and if technology can be used in conjunction with natural, social and human assets. The challenge now is to add the ecological dimension to crop productivity improvement. The term 'Evergreen Revolution' was coined about 15 years ago to indicate that we should develop technologies that can help to increase productivity without ecological harm. Productivity enhancement is the only pathway available to us to produce more to feed the growing population. Agro-ecosystems are to be manipulated to improve farming systems productivity on environmentally sustainable basis.

Conventional agriculture systems do not mimic natural systems and must be managed carefully with expensive inputs to maintain ecosystem health. Natural systems are more stable as nutrients are recycled, natural enemies prey on pests. The soil microorganisms enable decomposition to occur, providing a wealth of nutrients and organic matter to the system. If an agro-ecosystem is designed to mimic a natural system, we can minimize expensive inputs of fossil fuels and other limited resources. The challenge with such a system is to maintain a level of agricultural production efficient for ever increasing population while, still conserving resources. However, agroecosystems are most sustainable if they function as natural system. Crop ecology as a field of study looks at how the agroecosystem functions naturally and examines how adjustment can be made to the agroecosystem without destroying other facts of the system.

4.1. Principles and Basic Elements of Crop Ecology

Ecological principles

1. Enhance recycling of biomass and optimizing nutrient availability and balancing nutrient flow.
2. Securing favourable soil conditions for the plant growth particularly by managing organic matter and enhancing soil biotic activity.
3. Minimizing losses of carbon and energy due to flows of solar radiation, air and water by way of micro-climate management, water harvesting and soil management through increased soil cover.
4. Species and genetic diversification of the agroecosystem in time and space.
5. Enhance beneficial biological interactions and synergisms among agro biodiversity components. Thus, resulting in the promotion of the key ecological processes and mechanisms.

Ecological processes

1. Strengthen the immune system *i.e.*, proper functioning of natural pest control.
2. Decreasing toxicity levels through elimination of agrochemicals.
3. Optimize metabolic functions *i.e.*, organic matter decomposition and nutrient cycling.
4. Balance regulatory systems *i.e.*, nutrient cycles, water balance, energy flow, population control *etc.*
5. Enhance conservation and regeneration of soil-water resources and biodiversity.
6. Increase and sustain term productivity.

Mechanisms

1. Increase of plant species and genetic diversity in time and space
2. Enhancement of functional biodiversity *i.e.* natural enemies, antagonists *etc.*
3. Enhancement of soil organic matter and biological activity.

4. Increase soil cover and crop competition ability.
5. Elimination of toxic inputs and residues.

Basic concepts of ecology

1. The organisms and the environment reacting against each other form a complex unit which should be studied as a whole.
2. Organisms are elastic and get adapted in time and space to a range of environmental variations.
3. Organisms living together develop an economy of their own and organize into communities.
4. The environment is a complex of interrelated factors with temporal and spatial variations.
5. Plant activities such as dispersal, growth, competition, reproduction, death and decay lead to development of new vegetation and change in environmental characteristics. It leads to succession of plant communities.
6. Plant succession leads to development of a climax community which is relatively stable and is in equilibrium with the environment.

Agro-ecology: Agro-ecology is an integrated approach that simultaneously applies ecological and social concepts and principles to the design and management of food and agricultural systems. It seeks to optimize the interactions between plants, animals, humans and the environment while taking into consideration the social aspects that need to be addressed for a sustainable and fair food system.

Agro-ecology is not a new invention. It can be identified in scientific literature since 1920s, and has found expression in farmers' practices, in grassroots social movements for sustainability and the public policies of various countries around the world. More recently, agro-ecology has entered the discourse of international and UN institutions.

What makes agro-ecology distinct?

Agro-ecology is fundamentally different from other approaches to sustainable development. It is based on bottom-up and territorial processes, helping to deliver contextualised solutions to local problems. Agro-ecological innovations are based on the co-creation of knowledge, combining science with the traditional, practical and local knowledge of producers. By enhancing their autonomy and adaptive capacity, agro-ecology empowers producers and communities as key agents of change.

Rather than tweaking the practices of unsustainable agricultural systems, agro-ecology seeks to transform food and agricultural systems, addressing the root causes of problems in an integrated way and providing holistic and long-term solutions. This includes an explicit focus on social and economic dimensions of food systems. Agro-ecology places a strong focus on the rights of women, youth and indigenous peoples.

Elements of agro-ecology (FAO)

In guiding countries to transform their food and agricultural systems, to mainstream sustainable agriculture on a large scale, and to achieve Zero Hunger and multiple other Sustainable Development Goals, the following 10 elements emanated from the FAO regional seminars on agro-ecology.

» **Diversity; synergies; efficiency; resilience; recycling; co-creation and sharing of knowledge** (describing common characteristics of agro-ecological systems, foundational practices and innovation approaches)

» **Human and social values; culture and food traditions** (context features)

» **Responsible governance; circular and solidarity economy** (enabling environment)

The 10 elements of agro-ecology are interlinked and interdependent (**Fig**.6). As an analytical tool, the 10 elements can help countries to operationalise agro-ecology. By identifying important properties of agro-ecological systems and approaches, as well as key considerations in developing an enabling environment for agro-ecology, the 10 elements are a guide for policymakers, practitioners and stakeholders in planning, managing and evaluating agro-ecological transitions.

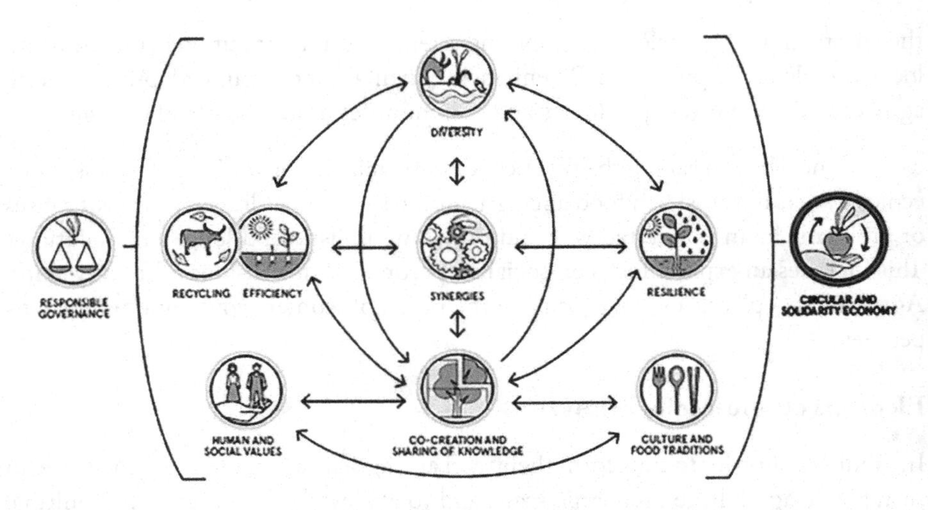

Fig.6. Elements of agro-ecology

1. **Diversity:** Diversification is key to agro-ecological transitions to ensure food and nutrition security while conserving, protecting and enhancing natural resources.

Agro-ecological systems are highly diverse. From a biological perspective, agro-ecological systems optimize the diversity of species and genetic resources in different ways. For example, agro-forestry systems organize crops, shrubs, and trees of different heights and shapes at different levels or strata, increasing vertical diversity. Intercropping combines complementary species to increase spatial diversity. Crop rotations, often including legumes, increase temporal diversity.

Crop–livestock systems rely on the diversity of local breeds adapted to specific environments. In the aquatic world, traditional fish polyculture farming, Integrated Multi-Trophic Aquaculture (IMTA) or rotational crop-fish systems follow the same principles to maximizing diversity.

Increasing biodiversity contributes to a range of production, socio-economic, nutrition and environmental benefits. By planning and managing diversity, agro-ecological approaches enhance the provisioning of ecosystem services, including pollination and soil health, upon which agricultural production depends. Diversification can increase productivity and resource-use efficiency by optimizing biomass and water harvesting.

Agro-ecological diversification also strengthens ecological and socio-economic resilience, including by creating new market opportunities. For example, crop and animal diversity reduces the risk of failure in the face of climate change. Mixed grazing by different species of ruminants reduces health risks from parasitism, while diverse local species or breeds have greater abilities to survive, produce and maintain reproduction levels in harsh environments. In turn, having a variety of income sources from differentiated and new markets, including diverse products, local food processing and agri-tourism, helps to stabilize household incomes. Consuming a diverse range of cereals, pulses, fruits, vegetables, and animal-source products contributes to improved nutritional outcomes. Moreover, the genetic diversity of different varieties, breeds and species is important in contributing macronutrients, micronutrients and other bioactive compounds to human diets. For example, in Micronesia, reintroducing an underutilized traditional variety of orange-fleshed banana with 50 times more beta-carotene than the widely available commercial white-fleshed banana proved instrumental in improving health and nutrition.

At the global level, three cereal crops provide close to 50 percent of all calories consumed, while the genetic diversity of crops, livestock, aquatic animals and trees continues to be rapidly lost.

Agro-ecology can help reverse these trends by managing and conserving agro-biodiversity, and responding to the increasing demand for a diversity of products that are eco-friendly. One such example is 'fish-friendly' rice produced from irrigated, rainfed and deepwater rice ecosystems, which values the diversity of aquatic species and their importance for rural livelihoods.

2. **Co-creation and sharing of knowledge:** Agricultural innovations respond better to local challenges when they are co-created through participatory processes. Agro-ecology depends on context-specific knowledge.

It does not offer fixed prescriptions – rather, agro-ecological practices are tailored to fit the environmental, social, economic, cultural and political context. The co-creation and sharing of knowledge plays a central role in the process of developing and implementing agro-ecological innovations to address challenges across food systems including adaptation to climate change. Through the co-creation process, agro-ecology blends traditional and indigenous knowledge, producers and traders practical knowledge, and global scientific knowledge. Producer's knowledge of agricultural biodiversity and management experience for specific contexts as well as their knowledge related to markets and institutions are absolutely central in this process.

Education - both formal and non-formal plays a fundamental role in sharing agro-ecological innovations resulting from co-creation processes. For example, for more than 30 years, the horizontal *campesino a campesino* movement has played a pivotal role in sharing agro-ecological knowledge, connecting hundreds of thousands of producers in Latin America. In contrast, top-down models of technology transfer have had limited success. Promoting participatory processes and institutional innovations that build mutual trust enables the co-creation and sharing of knowledge, contributing to relevant and inclusive agro-ecology transition processes.

3. **Synergies:** Building synergies enhances key functions across food systems, supporting production and multiple ecosystem services.

Agro-ecology pays careful attention to the design of diversified systems that selectively combine annual and perennial crops, livestock and aquatic animals, trees, soils, water and other components on farms and agricultural landscapes to enhance synergies in the context of an increasingly changing climate.

Building synergies in food systems delivers multiple benefits. By optimizing biological synergies, agro-ecological practices enhance ecological functions, leading to greater resource-use efficiency and resilience. For example, globally biological nitrogen fixation by pulses in intercropping systems or rotations generates close to USD 10 million savings in nitrogen fertilizers every year, while contributing to soil health, climate change mitigation and adaptation. Furthermore, about 15 per cent of the nitrogen applied to crops comes from livestock manure, highlighting synergies resulting from crop–livestock integration. In Asia, integrated rice systems combine rice cultivation with the generation of other products such as fish, ducks and trees. By maximising synergies, integrated rice systems significantly improve yields, dietary diversity, weed control, soil structure and fertility, as well as providing biodiversity habitat and pest control.

At the landscape level, synchronization of productive activities in time and space is necessary to enhance synergies. Soil erosion control using *Calliandra* hedgerows is common in integrated agro-ecological systems in the East African Highlands. In this example, the management practice of periodic pruning reduces tree competition with crops grown between hedgerows and at the same time provides feed for animals, creating synergies between the different components. Pastoralism and extensive livestock grazing systems manage complex interactions between people, multi-species herds and variable environmental conditions, building resilience and contributing to ecosystem services such as seed dispersal, habitat preservation and soil fertility.

While, agro-ecological approaches strive to maximize synergies, trade-offs also occur in natural and human systems. For example, the allocation of resource use or access rights often involve trade-offs. To promote synergies within the wider food system, and best manage trade-offs, agro-ecology emphasizes the importance of partnerships, cooperation and responsible governance, involving different actors at multiple scales.

4. **Efficiency:** Innovative agro-ecological practices produce more using less external resources their by improves resource use efficiency.

Increased resource use efficiency is an emergent property of agro-ecological systems that carefully plan and manage diversity to create synergies between different system components. For example, a key efficiency challenge is that less than 50 per cent of nitrogen fertilizer added globally to crop land is converted into harvested products and the rest is lost to the environment causing major environmental problems.

Agro ecological systems improve the use of natural resources, especially those that are abundant and free, such as solar radiation, atmospheric carbon and nitrogen. By enhancing biological processes and recycling biomass, nutrients and water, producers are able to use fewer external resources, reducing costs and the negative environmental impacts of their use. Ultimately, reducing dependency on external resources empowers producers by increasing their autonomy and resilience to natural or economic shocks.

One way to measure the efficiency of integrated systems is by using Land Equivalent Ratios (LER). LER compares the yields from growing two or more components (e.g. crops, trees, animals) together with yields from growing the same components individually. Integrated agro-ecological systems frequently demonstrate higher LERs. Agro-ecology, thus promotes agricultural systems with the necessary biological, socio-economic and institutional diversity and alignment in time and space to support greater efficiency.

5. **Recycling:** More recycling means agricultural production with lower economic and environmental costs.

Waste is a human concept – it does not exist in natural ecosystems. By imitating natural ecosystems, agro-ecological practices support biological processes that drive the recycling of nutrients, biomass and water within production systems, thereby increasing resource use efficiency and minimizing waste and pollution.

Recycling can take place at both farm-scale and within landscapes, through diversification and building of synergies between different components and activities.

For example, agro-forestry systems that include deep rooting trees can capture nutrients lost beyond the roots of annual crops. Crop–livestock systems promote recycling of organic materials by using manure for composting or directly as fertilizer, and crop residues and by-products as livestock feed.

Nutrient cycling accounts for 51 per cent of the economic value of all non-provisioning ecosystem services, and integrating livestock plays a large role in this. Similarly, in rice–fish systems, aquatic animals help to fertilize the rice crop and reduce pests, reducing the need for external fertilizer or pesticide inputs.

Recycling delivers multiple benefits by closing nutrient cycles and reducing waste that translates into lower dependency on external resources, increasing the autonomy of producers and reducing their vulnerability to market and climate shocks. Recycling organic materials and by-products offers great potential for agro-ecological innovations.

6. **Resilience:** Enhanced resilience of people, communities and ecosystems is the key to sustainable food and agricultural systems.

Diversified agro-ecological systems are more resilient as they have a greater capacity to recover from disturbances including extreme weather events such as drought, floods or hurricanes, and to resist pest and disease attack. Following Hurricane Mitch in Central America in 1998, biodiverse farms including agro-forestry, contour farming and cover cropping retained 20–40 per cent more topsoil, suffered less erosion and experienced lower economic losses than neighboring farms practicing conventional monocultures.

By maintaining a functional balance, agro-ecological systems are better able to resist pest and disease attack. Agro-ecological practices recover the biological complexity of agricultural systems and promote the necessary community of interacting organisms to self regulate pest outbreaks.

On a landscape scale, diversified agricultural landscapes have a greater potential to contribute to pest and disease control functions. Agro-ecological approaches can equally enhance socio-economic resilience. Through diversification and integration, producers reduce their vulnerability should a single crop, livestock species or other commodity fail.

By reducing dependence on external inputs, agro-ecology can reduce producers' vulnerability to economic risk. Enhancing ecological and socioeconomic resilience go hand-in-hand after all, humans are an integral part of ecosystems.

7. **Human and social values:** Protecting and improving rural livelihoods, equity and social well-being is essential for sustainable food and agricultural systems.

Agro-ecology places a strong emphasis on human and social values, such as dignity, equity, inclusion and justice all contributing to the improved livelihoods dimension of the SDGs. It puts the aspirations and needs of those who produce, distribute and consume food at the heart of food systems. By building autonomy and adaptive capacities to manage their agro-ecosystems, agro-ecological approaches empower people and communities to overcome poverty, hunger and malnutrition, while promoting human rights, such as the right to food, and stewardship of the environment so that future generations can also live in prosperity.

Agro-ecology seeks to address gender inequalities by creating opportunities for women. Globally, women make up almost half of the agricultural workforce. They also play a vital role in household food security, dietary diversity and health, as well as in the conservation and sustainable use of biological diversity. In spite of this, women remain economically marginalized and vulnerable to violations of their rights, while their contributions often remain unrecognized. Agro-ecology can help rural women in family farming agriculture to develop higher levels of autonomy by building knowledge, through collective action and creating opportunities for commercialization. Agro-ecology can open spaces for women to become more autonomous and empower them at household, community levels and beyond – for instance, through participation in producer groups. Women's participation is essential for agro-ecology and women are frequently the leaders of agro-ecology projects.

In many places around the world, rural youth face a crisis of employment. Agro-ecology provides a promising solution as a source of decent jobs. Agro-ecology is based on a innovative way of agricultural production that is knowledge intensive, environmentally friendly, socially responsible, innovative, and which depends on skilled labours. Meanwhile, rural youth around the world possess energy, creativity and a desire to positively change their world. What they need is support and opportunities. As a bottom-up, grassroots paradigm for sustainable rural development, agro-ecology empowers people to become their own agents of change.

8. **Culture and food traditions:** By supporting healthy, balanced, diversified and culturally appropriate diets, agro-ecology contributes to food security and nutrition while maintaining the health of ecosystems.

Agriculture and food are core components of human heritage. Hence, culture and food traditions play a central role in society and in shaping human behavior.

However, in many instances, our current food systems have created a disconnection between food habits and culture. This disconnection has contributed to a situation where hunger and obesity exist side by side, in a world that produces enough food to feed its entire population.

Almost 800 million people worldwide are chronically hungry and 2 billion suffer micronutrient deficiencies. Meanwhile, there has been a rampant rise in obesity and diet-related diseases; 1.9 billion people are overweight or obese and non-communicable diseases (cancer, cardiovascular disease, diabetes) are the number one cause of global mortality. To address the imbalances in our food systems and move towards a zero hunger world, increasing production alone is not sufficient.

Agro-ecology plays an important role in re-balancing tradition and modern food habits, bringing them together in a harmonious way that promotes healthy food production and consumption, supporting the right to adequate food. In this way, agro-ecology seeks to cultivate a healthy relationship between people and food. Cultural identity and sense of place are often closely tied to landscapes and food systems. As people and ecosystems have evolved together, cultural practices and indigenous and traditional knowledge offer a wealth of experience that can inspire agro-ecological solutions. For example, India is home to an estimated 50000 indigenous varieties of rice bred over centuries for their specific taste, nutrition and pest resistance properties, and their adaptability to a range of conditions. Culinary traditions are built around these different varieties, making use of their different properties. Taking this accumulated body of traditional knowledge as a guide, agro-ecology can help realize the potential of territories to sustain their peoples.

9. **Responsible governance:** Sustainable food and agriculture requires responsible and effective governance mechanisms at different scales – from local to national to global.

Agro-ecology calls for responsible and effective governance to support the transition to sustainable food and agricultural systems. Transparent, accountable and inclusive governance mechanisms are necessary to create an enabling environment that supports producers to transform their systems following agro-ecological concepts and practices. Successful examples include school feeding and public procurement programmes, market regulations allowing for branding of differentiated agro-ecological produce, and subsidies and incentives for ecosystem services.

Land and natural resources governance is a prime example. The majority of the world's rural poor and vulnerable populations heavily relies on terrestrial and aquatic biodiversity and ecosystem services for their livelihoods, yet lack secure access to these resources. Agro-ecology depends on equitable access to land and natural resources – a key to social justice, but also in providing incentives for the long-term investments that are necessary to protect soil, biodiversity and ecosystem services.

Agro-ecology is best supported by responsible governance mechanisms at different scales. Many countries have already developed national level legislation, policies and programmes that reward agricultural management that enhances biodiversity and the provision of ecosystem services. Territorial, landscape and community level governance, such as traditional and customary governance models, is also extremely important to foster cooperation between stakeholders, maximizing synergies while reducing or managing trade-offs.

10. **Circular and solidarity economy:** Circular and solidarity economies that reconnect producers and consumers provide innovative solutions for living within our planetary boundaries while, ensuring the social foundation for inclusive and sustainable development.

Agro-ecology seeks to reconnect producers and consumers through a circular and solidarity economy that prioritizes local markets and supports local economic development by creating virtuous cycles. Agro-ecological approaches promote fair solutions based on local needs, resources and capacities, creating more equitable and sustainable markets. Strengthening short food circuits can increase the incomes of food producers while maintaining a fair price for consumers. These include new innovative markets, alongside more traditional territorial markets, where most smallholders market their products.

Social and institutional innovations play a key role in encouraging agro-ecological production and consumption. Examples of innovations that help link producers and consumers include participatory guarantee schemes, local producer's markets, denomination of origin labeling, community supported agriculture and e-commerce schemes. These innovative markets respond to a growing demand from consumers for healthier diets.

Re-designing food systems based on the principles of circular economy can help to address the global food waste challenge by making food value chains shorter and more resource-efficient. Currently, one third of all food produced is lost or wasted, failing to contribute to food security and nutrition, while exacerbating pressure on natural resources. The energy used to produce food that is lost or wasted is approximately 10 per cent of the world's total energy consumption, while the food waste footprint is equivalent to 3.5 Gt CO_2 of greenhouse gas emissions per year.

Chapter - 5

Plant Distribution and Adaptations

Study of ecology provides us answers to why and how an organism survives in a habitat. How and from where the organism come to occupy the habitat and get established there, is answered through a study of bio-geography. Bio-geography is thus the study of present and past geographical distribution of organisms on the earth. Study of migration and distribution of animals is known as **Zoo-geography,** while that of plants is known as **Plant geography** or **Phyto-geography.**

Phyto-geography (Greek word, *phyton* = "plant" and *geographia* = "geography" meaning distribution) or botanical geography is the branch of biogeography that is concerned with the geographic distribution of plant species and their influence on the earth's surface. Phyto-geography is concerned with all aspects of plant distribution, from the controls on the distribution of individual species ranges (at both large and small scales, see species distribution) to the factors that govern the composition of entire communities and floras.

Prussian Naturalist Alexander Van Humbolt (1806) is called as 'Father of plant geography'. He detailed about plant and animal distribution in his book "Plant geography"

According to Campbell (1926), the main theme of plant geography is to discover the similarities and diversities in the plants and floras of the present and past found in widely separated parts of the earth.

Wulff (1943) states that Phyto-geography is the study of distribution of plant species in their habitats and elucidation of origin and history of development of floras.

According to Croizat (1952), Phyto-geography is the study of migration and evolution of plants in time and space.

Major divisions of Phyto-geography:

There are two major divisions of Phyto-geography:

(i) Descriptive or Static Phyto-geography

(ii) Interpretive or Dynamic Phyto-geography

Descriptive phyto-geography: This deals with the actual description of floristic or vegetational groups found in different parts of the world. Early plant geographers described floras and attempted to divide earth into floristic and botanical zones.

Interpretive or dynamic phyto-geography: This deals with the dynamics of migration and evolution of plants and floras. It explains the reasons for varied distribution of plant species in different parts of the world. It is a borderline science involving synthesis and integration of data and concepts from several specialized disciplines like ecology, physiology, genetics, taxonomy, evolution, palaeontology and geology. Good (1931), Mason (1936), Cain (1944) and some others have pointed out the factors involved in the distribution of plants.

Good (1931) another plant ecologist given principles of plant distribution which as follows:

1. **The plant distribution is primarily controlled by distribution of climatic conditions:** This principle is expressed in the general world-wide correlation between climatic conditions and vegetation. It is seen most conspicuously in the parallel between the main temperature and floral zonations of the world, and in the segregation of the plant life of individual temperature zones according to the distribution of the rainfall. It is epitomised in the altitudinal zonation of the vegetation in elevated regions, and the distribution of crop and garden plants also illustrates it.

2. **The plant distribution is secondarily controlled by distribution of edaphic factors:** The edaphic factors are the soil properties that affect the diversity of organisms living in the soil environment. These include soil structure, temperature, pH, and salinity. Some of them are influenced by man, but most are independent of human activity. Few plants are able to exist in more than a few kinds of habitat and most are confined more or less rigidly to one or two. It follows that while

a species may be expected to occur, as far as climate is concerned, over a wide area, its actual range will depend upon the distribution of appropriate edaphic conditions within this area. It will tend to be found, within its climatic area, wherever there are suitable habitats, but it cannot exist outside its climatic area whatever the edaphic conditions may be. The importance of edaphic factors as determining plant range is usually subordinate to that of climatic factors, because, normally, the distribution of edaphic conditions is much more detailed and irregular than that of climatic factors and has no fundamental zonation corresponding to that of the world temperature gradient. It is rare to find a homogeneous climatic area which is not, edaphically, heterogeneous.

3. **The great movements of floras have taken place in the past and are still continuing:** This principle concerns the movement of species or assemblies of species as opposed to the movement of individual plants. Movement of this kind, conveniently called plant migration, is indicated wherever there is, in one locality, evidence of a succession of different floras, because the replacement of one flora by another must inimitably mean considerable lateral floral movement. The fossil record affords innumerable instances of these successions at all times from the earliest to the present, and there is no reason to support that, the process has now ceased.

4. **That species movement (plant migration) is brought about by the transport of individual plants during their motile dispersal phases:** No plant which live rooted in soil are capable of natural transport without considerable risk and a strong probability of destruction. They are motile only in the form of reproductive units and effective transport can take place only when they are in this condition. In certain free-floating aquatics the whole or part of the plant serve functions as a dispersal unit.

5. **There has been great variation and oscillation in climate, especially at higher latitudes, during the geological history of the Angiosperms:** The Angiosperms, as we know the group today, may, for present purposes, be considered as dating from the early part of the Cretaceous period. There is abundant evidence of climatic change since that time. Geological evidence is two-fold, the occurrence of certain rocks whose formation or modification is associated with particular climatic conditions, and the phenomena of glaciation as seen in the Pleistocene. Botanical evidence is afforded by the fossil record, and by the study of the growth rings of long-lived trees like the Sequoias. Archaeologieal evidence of minor climatic changes during the human period is also plentiful.

6. **Variation has occurred in the relative distribution and outline of land and sea during the history of the Angiosperms:** The fact that much of the present land surface of the globe is made up of rocks originally deposited beneath the sea is sufficient basis for the general statement of geographical change, but it is believed that the changes have been greater than is indicated simply by these marine beds. In particular some connection in the past between the present continents is usually thought necessary to explain the facts of animal and plant distribution. Two theories are put forward to supply this connection. The theory of 'land bridges' supposes that the continents have remained *in-situ* but that from time to time in the past there have existed connecting land masses between them. The theory of 'continental displacement' on the other hand supposes that the present continents were once all joined into one super continental land mass, and have attained their present positions and shapes by the disruption of this mass and the divergent drift of the constituent fragments.

Lowerence (1951) has suggested the following thirteen modern principles of Phyto-geography which are classified into four groups:

I. Principles concerning environment:

1. The distribution of plants is primarily controlled by climatic conditions.
2. There has been variation in climate during geological history in the past which affected migration of plants.
3. The relations between land masses and seas have varied in the past. The large land masses split up to form new land masses or continents which separated and reoriented. Land bridges between continents acted as probable routes for migration of plant and animal species. The land bridges became submerged in sea with the passage of time and the possibility for migration of plants and animals from one continent to another disappeared for ever.
4. Soil conditions on plains and mountains of different land masses show secondary control on distribution of vegetation. Halophytes, psammophytes, calcicols, calcifobs etc. have developed because of edaphic conditions.
5. Biotic factors also play important role in distribution and establishment of plant species.
6. The environment is holocentric, *i.e.*, all environmental factors have combined effects on the vegetation of a place (Ale and Pank, 1939).

II. Principles concerning plant responses:

7. Range of distribution of plants is limited by their tolerances. Each plant species has a range of climatic and edaphic conditions. Therefore, tolerance of a large taxon is the sum of tolerances of its constituent species.

8. Tolerances have a Genetic basis. The response of plants to environment is governed by their genetic makeup. Many of the crops through breeding and genetic changes have been made to grow in wider range of environmental conditions. In nature, hybrid plants have been found to have wider range of tolerances than their parents.

9. Different ontogentic phases have different tolerances. Different developmental stages of plants show different degree of tolerances, as for example seeds and mature plants are more tolerants to temperature and moisture variations than their seedlings.

III. Principles concerning the migration of floras and climaxes:

10. Large scale migrations have taken place. The fossils and palaeoecological evidences reveal that large scale migrations of plants and animals have taken place during Mesozoic era and Tertiary periods.

11. Migration resulted from transport and establishment. In the process of migration plants are dispersed to new habitats through their propagules such as spores, seeds, bulbils etc., and there they are established if environmental conditions are favourable. Plants grow and reproduce there and progeny perpetuates through ecological adjustments.

IV. Principles concerning the perpetuation and evolution of floras and climaxes:

12. Perpetuation depends first upon migration and secondly upon the ability of species to transmit the favourable variations to the progenies.

13. Evolution of floras and climaxes depends upon migration, evolution of species and environmental selections.

Tolerance theory: Tolerance is the ability of an organism to withstand and reproduce under a definite standard of conditions or ranges. There are mainly two theories of tolerance and they are 'General theory of tolerance' proposed by American zoologist Victor Ernest Shelford **in 1911** and 'Specific theory of tolerance' proposed by Good.

General theory of tolerance: Each and every plant species is able to exist and reproduce successfully only within a definite range of climatic and edaphic conditions. This range represents the tolerance of the species to external conditions. The tolerance of any species is a specific character subject to the laws and processes of organic evolution in the same way as its morphological characters, but the two are not necessarily linked. Change in tolerance may or may not be accompanied by morphological change and morphological change may or may not be accompanied by change of tolerance. Morphologically similar species may show wide differences in tolerance and species with similar tolerance may show very little morphological similarity. The relative distribution of species with similar ranges of tolerance is finally determined by the result of the competition between them. The tolerance of any larger taxonomic unit is the sum of the tolerances of its constituent species.

Specific theory of tolerance: The further elaboration on the theory of tolerance is credited to Ronald Good who included the following concepts in specific theory of tolerance.

1. Organism with wide range of tolerance for all factors of the environment are likely to be widely distributed
2. Organisms may have a wide range of tolerance for one factor and a narrow range for another.
3. When conditions are not optimum for one factor, the limits of tolerance may be reduced with respect to another factor.
4. The period of reproduction usually is critical, when environmental factors are likely to be limiting.

Limiting factors: The chemist, Leibig (1840) expressed his Law of minimum which states that growth is dependent/limited by the pace of the slowest factor.

Distribution: On the basis of area of the earth surface occupied by the plants, the various taxa are categorized as under:

1. Wides
2. Endemics
3. Discontinuous species

1. Wides: Plants widely distributed over the earth in definite climatic zones and the different continents are referred to as wides. Cosmopolitan is applied for wides but, in fact, no plant is cosmopolitan in real sense of the term. *Taraxacum officinale* and *Chaenopodium album* are the common examples of the wides. Plants of tropical regions are called Pantropical. The plants of very cold climate may not only be found in the arctic regions but also in alpine zone of mountains in tropical and subtropical regions. These are called arctic-alpine plants.

2. Endemics: A taxon whose distribution is confined to a given area is said to be endemic to that area. The taxon may be of any rank, although it is usually at a family level or below, and its range of distribution may be wide, spanning an entire continent, or very narrow covering only a few square metres. The concept of endemism is important because in the past the formulation of bio-geographic regions was based on it.

The limits of a region are determined by mapping the distributions of taxa, where the outer boundaries of many taxa occur, a line delimiting a bio-geographic region is drawn. Major regions are still determined as those that have the most endemics or stated another way, those that share the fewest taxa with other regions. As regions are further broken down into subdivisions, they will contain fewer unique taxa.

This has been criticized because it assumes that species ranges are stable, which they are not. An alternative method of determining bio-geographic regions involves calculating degrees of similarity between geographic regions. The concept of endemic distribution of plants was put forth by A.P. De Candolle (1813). Engler (1882) suggested two categories of endemic forms; Palaeo-endemics which are survivors of ancient forms and indigenous or native forms which are confined to a particular ocahty. According to area of distribution, the species may be continental endemics (restricted to a continent, endemic to a country, provincial, regional) or local endemics (restricted to valley, hills, islands, etc.).

Now the endemic species have been grouped into the following categories:

(i) Relics or Palaeoendemics: They are the survivors of once widely distributed ancestral forms, for example, *Ginkgo biloba* (restricted to China and Japan), *Sequoia sempervirens* (confined to coastal valleys of California, U.S.A.). *Agathis australis, Metasequoia* (Confined to Single valley in China). These species are called Palaeo-endemics or epibionts. A great majority of the endemic species belonging to this type have many fossil relatives. They are also called living fossils. Because of little variability the endemics are adapted only to a particular environment and even if they reach new areas, they fail to establish themselves in new environment.

(ii) Neo-endemics: The other endemics may be modem species which have had not enough time for occupying a large area through migration. They are called neo-endemics. There are several such genera which are widely endemic or few species of which are endemic. Neo-endemics show good variability and have many biotypes, grow in diverse habitats and have wide tolerance for habitats.

Some of the well known endemic genera in Indian flora are *Mecanopsis* (Papaveraceae) *Chloroxylon swietenia* (Flindersiaceae, formerly Rutaceae). *Catenaria* and Butea (Papilionaceae) *Caesulia* (Compositae), *Petalidium* (Acanthaceae), *etc. Eletteria repens* (Zingiberaceae) *Piper longum* (Piperaceae), *Piper nigrum* (Piperaceae), *Ficus religiosa* (Moraceae), *Shorea robusta* (Dipterocarpaceae), *Venda caerulea* (Orchidaceae), *Salmalia malabarica* (Bombacaceae) *Eleusine coracana* (Grammeae) are the well known endemic species of Indian flora.

There are some special terms to designate the quality of these endemics, *viz.*, Local endemics which are found in small land features, progressive endemics which tend to spread with time retrogressive endemics in which case the area of distribution is contracting and micro-endemics (*i.e.*, the endemics of lower groups).

(iii) Pseudo endemics: These endemics arise due to mutation in existing population at a particular place. These pseudo endemics or mutants may or may not persist for long in the particular area where they originate. Endemism results from the failure on the part of species to disseminate its seeds fruits spores or propagules because of existence of great barriers like mountains, oceans and large deserts. The oceanic islands which are isolated from rest of the world by large expanses of water abound in endemic species and water barrier checks the migration of those species outside their original habitat.

3. Discontinuous distribution: When plants occur at two or more distant places of the world which are separated by overland's or oceans hundreds or thousands of kilometres apart. Such a distribution is called discontinuous or disjunct distribution. The genera *Nothofagus, Jovellona* are good examples and they are found in parts of South America, South Africa and Australia which are paraded by vast oceans.

The significant phyto-geographical causes for discontinuous distribution are as follows:

i. The species might have evolved at more than one place and they failed to migrate outside their original habitats because of barriers.

ii. The species which were once widely distributed in the past disappeared from certain areas and are now surviving in some distant pockets.

iii. The climate may also be a factor for discontinuity in distribution of species. Plants having specific climatic requirements are found in widely separated areas with similar environmental conditions, as for example, plants of arctic regions are also found in alpine zone of high mountains in tropics and subtropics. *Salix* and *Silen* species show discontinuous distribution in arctic-alpine regions.

Theories of discontinuous distribution

1. Theory of land bridge: According to this theory, land bridges occurring in between the separated continents are believed to have helped in the migration of various taxa from one continent to the other. Uniform distribution of plants and animals in different parts of the world during Palaeozoic era is believed to have been due to those land bridges. With the passage of time the land bridges became submerged in sea and the connections between the various continents snapped beyond the dispersal capacity of organisms resulting thereby the discontinuity in the distribution.

2. Theory of continental drift: The theory of continental drift was propounded by Wegner (1912-1924). According to him the whole land mass of the world was a single super continent during Palaeozoic era. He named it as Pangaea. That super continent was surrounded by sea on all the sides which was named Panthalassa. During Mesozoic era, Pangaea split up into two large landmasses, Laurasia in the north and Gondwanaland in south.

The two landmasses were separated by Tethys Sea. Du Toit (1937), however, suggested that Laurasia and Gondwanaland existed from the very beginning. The two large landmasses having characteristic flora and fauna broke up into new landmasses called continents. Laurasia gave rise to Eurasia, Greenland and North America and similarly Gondwanaland gave rise to South America, Africa, India, and Polynesia, Australia Antarctica etc.

About 135 million years ago reorientation of continents began. The continents were drifted apart by the oceans. This is called Continental Drift. The occurrence of Dinosaurs and many fossil plants lend support to the existence of Laurasia and Gondwanaland. With the separation of continents the distribution areas of several plant and animal species got separated and gave rise to discontinuous distribution areas.

Factors affecting distribution of species

Several factors are known to affect the geographical distribution of plant species, some of which are as follows:

1. Geological history and distribution
2. Migration and
3. Ecological amplitude.

1. Geological history and distribution: The place where a species first originated is called its centre of origin. Evolution of species is a slow but continuous process. Some of the species in present day flora are quite old while a great majority of them are recent in origin.

The process of species differentiation involves:

(i) Hybridization between the related species as well as mutation

(ii) The natural selection of the hybrid and mutant populations.

In the selection process not all the hybrids and mutants are selected by nature and only the fittest individuals which find the habitat conditions within their ecological amplitudes are selected and the individuals least fit are eliminated. Changing climate has also played important role in the origin of new species. In the course of evolution several old species became extinct, some of which can be found even today as fossils. The fossils provide direct evidence for the existence of various taxa in the past.

Age and area hypothesis:

This hypothesis was proposed by **J. W. Willis** (1915) on the basis of his extensive studies of geographical distribution of certain plant species in tropics. On the basis of his findings Willis postulated that the species which evolved earlier occupy greater areas than those which appeared later in the evolutionary sequence. According to this hypothesis, the frequency of a species over an area is directly proportional to its age in scale of evolution and age of species is directly related with the area of its distribution.

Thus a small area of distribution of a species will indicate its relative young age. Willis has quoted several examples such as *Impatiens, Primula, Gentiana, Rhododendron* in support of his hypothesis. Genus *Coleus* may be quoted here as an example in support of this hypothesis and the two species of Coleus namely *C. elongatus* and *C. barbatus*. The former species is endemic while the latter is widely distributed. On the basis of areas under distribution of these species Willis considered *C. elongatus* less evolved and derived from *C. barbalins*. Willis has also pointed out that the majority of endemics are found to be members of large and successful genera. The age and area hypothesis, however, is not universal and it has been criticized by many.

2. Migration: The newly evolved species starts migration to new areas and side by side it undergoes further evolutionary changes. The dispersal of germules and propagules is brought about by several agencies like wind, water, glaciers, insects, animals and even man. The dispersal is followed by ecasis. Migration may be adversely affected and sometimes even totally stopped by some factors called migration barriers. Barriers in the dispersal of species may be classified as ecological or environmental and geographical.

The climate, an ecological barrier, plays important role in distribution and establishment of species. Unsuitable climatic condition or change of climate in particular area forces the species to migrate from one place to another and the failure of some species to migration leads them to gradual extinction. Besides climate, there are geographical barriers too, for example, high mountains, vast oceans or deserts.

The fresh water plants, cannot be dispersed across oceans if their propagules are suitable only for fresh water dispersal and similarly germules or propagules of land plants from one country cannot reach other country separated by vast oceans and mountains. Species are called natives of the place of occurrence if they originated there. Outside the area of its origin, the species is referred to as exotic. Exotic species reach new area through migration. If any species is introduced intentionally in new area by man then it is called introduced species.

3. Ecological amplitudes and distribution: Environmental conditions not only influence the life and development of plants but also determine the presence or absence, vigour or weakness and relative success or failure of various plants in a particular habitat. Each plant species of a community has a definite range of tolerance towards physical and biological environment of the habitat. This is referred to as ecological amplitude. The presence of species at a particular place, no doubt, indicates that the environmental conditions of that habitat are within its ecological amplitude but the absence of a species from one place does not necessarily indicate that the environment is not suitable for that species.

The ecological amplitude is governed by genetic set up of the species concerned and thus different species have different ecological amplitudes which may sometimes overlap only in certain respects. Further, some species may occur at different geographical regions as and when the conditions fall within their ecological amplitudes. As for example, some plants of temperate region say conifers, may also be found in alpine zone of high mountains in tropical and subtropical regions.

The other consideration in ecological amplitude as a factor in plant distribution is its change with time. In sexually reproducing plants the hybridization between related species results in offspring's with new genetic composition. With the change of environment the plant species also make adjustments with new environment by shifts in their ecological amplitudes facilitated by changes in the genotype. Within a species there may occur several genetically different groups of individuals (populations) which are adjusted to particular set of ecological conditions.

These populations are called ecotypes or ecological races or ecological populations. In *Euphorbia thymifolia*, for example, there are two major populations one is calcium loving or calcicole and the other type is calcium hating or calcifuge. Similarly ecological races of *Xanthium strumarium* and *Ageratum conyzoides* differ in the photoperiodic requirements. The existence of ecotypes within the species widens the area of its geographical distribution.

Apart from these factors, different plant geographers had different opinions on the factors affecting plant distribution. **Tansley** (1923) recognized four group of factors that affect the plant distribution:

1. Climatic
2. Physiographic
3. Edaphic
4. Biotic factors

According to **Billings** (1952), environment includes all the external forces and substances effecting the growth, structure and reproduction of plant. Billings subdivided environment into five large groups of factors: climatic, edaphic, geographic, pyric and biotic factors.

Basic principles and concepts relating to the natural distribution of plants through the world are useful in their application to crop adaptation. The distribution of plants has resulted from both natural causes and the activities of man. Man's activities have resulted in a widespread distribution of crop species. Some of the crops distribution as influenced by man activities are detailed below.

Banana: It is adapted to low elevation, well drained lands in the tropics, where no frost occurs. Uniformly high temperature with maximum sunshine is essential for high yields. Optimum temperature ranges from 23.9 to 29.4 °C with a minimum tolerance of 7.2 to 8.8 °C. Large quantities of water are required for successful production, preferably 1875 to 2500 mm rainfall so distributed that there is no drought.

Soils should be fertile, friable and have good aggregation. The top soil is particularly important because banana has a shallow root system. Clay content should be less than 40 %. Good drainage is absolutely necessary for the production of high quality fruit. The banana is best adapted to locations having little or no wind. When the fruit is maturing, wind damage may be disastrous as plants may be blown down and much of the crop is lost. Prevalence of panama disease changed the production picture of banana. Some areas have been completely abandoned.

Sugarcane: sugarcane grows best in frost free climate with warm to hot sunny weather. Such conditions are found in Cuba, Puerto Rico and Hawaii, where the monthly mean temperature ranges from 21.1 to 26.7 °C throughout the year. A root temperature of 26.7 °C is optimum for growth of cane and nutrient absorption. At below 21.1 °C, root growth is retarded and 10 °C no growth takes place. Sugarcane is grown from 37° N in Southern Spain to 31° S in South Africa. Optimal performance is observed around 20° latitude. In India, it is grown from 8-33° N and mostly from sea level to 1500m.

In the sub-tropics, sugarcane is grown at latitude of 30°. Growth and yields are limited by winter temperature. Important sugarcane producing countries include Brazil, Cuba, Pakistan, Thailand, Philippines, Argentina, Colombia, Indonesia, South Africa and Egypt besides India. Sugarcane cannot tolerate freezing temperature and growth ceases at mean temperature below 12 °C.

Coffee: Coffee is grown from sea level to 7000 ft. Altitude is important chiefly in its effect on temperature, rainfall and humidity, but it is also believed that the lower atmospheric pressure at higher altitudes also may have a stimulating effect. Coffee Arabica is best adapted to higher altitudes. Gentle slopes and adequate soil and air drainage are necessary. Coffee trees soon die in flat areas on heavy clay soils. This adaptation has restricted coffee growing mainly to the hilly or rolling uplands.

Maize: It is essentially a crop of warm countries with adequate moisture. Bulk of the crop is grown in warmer parts of temperate regions and in humid subtropics. It is mainly grown from 50° N to 40° S. It cannot withstand frost at any stage of its growth. It can be successfully grown where the night temperature does not go below 15.6 °C. Silking and tasseling stages of maize are adversely affected by high temperature (37.8 °C or more). At high temperatures, the internal water supply will be too low for pollen germination. One of the greatest limiting factors in corn yields is insufficient moisture and nutrient supply. The latitude of maize growing areas in India ranges from 12 to 30° N. The range in altitude is from 49 m in Bihar to 1250 m Himachal Pradesh.

Cotton: It is adapted to the humid sub-tropics, where the frost free season is at least 200-210 days. It is grown from 32° S to 37° N. Too much rain causes excessive shedding of leaves, squares, blooms and bolls.

The distribution of different crops is influenced by photoperiodic responses. Corn or sorghum from the tropics seldom mature when planted in regions where the days are long. Soil texture has an important influence on crop adaptation. Medium or heavy soils are best for fine rooted grasses, wheat and oats whereas, rye, corn and sorghum plants can thrive on the light sandy soils. Rice demands heavy soil with nearly impervious subsoil.

Distribution of major crop groups in the world

Cereals: The most prevalent group of crops across the world is cereals. Taken together, cereals are the only group of crops with cultivation that exceeds 20% of global land area or 61% of the total cultivated land. In particular, wheat, maize, barley, rice and millet are dominant and occupy two thirds of the cropland of the world.

Table 1: Area and relative proportion of crops in the world

Crop group	Area (1000 km^2)	Percentage of cultivated land
Cereals	10955	61
Roots and tubers	734	4
Sugar crops	419	2
Pulses	794	4
Oil crops	1819	10
Fiber crops	534	3
Others	2664	15
Total cropland	17920	100

Wheat is the most abundant crop occupying 22% of the cultivated area in the world. The most intensive wheat cultivation occurs in the temperate latitudes of both hemispheres. Wheat is most prevalent in the great plains of the USA, the Canadian prairie provinces, the Indus and the upper Ganges valleys, the Kazakhistan and Russian border and in southern Australia. Wheat is also found throughout Europe, in Southern South America, in parts of Eastern Africa and in Eastern China.

Barley and rye are preferentially grown in colder latitudes with the majority being cultivated around 55° N in Canada, the Northern US and European Russia.

Rice, sorghum and millets dominate the tropical and subtropical belts, especially in the Northern Hemisphere. These three crops occupy 11%, 3% and 2%, respectively of the global cultivated area. Rice is the second most extensive crop in the world and is the major crop of South and Southeast Asia. It is also cultivated in the Amazon basin, the Southern US and Southern Australia.

Sorghum, the only cereal that does not emerge as a dominant crop in any region is common throughout the Rift valley and the Sahel region in Africa, the Southern half of the Mississippi valley and India. Over 90% of Indian sorghum is grown between 12 and 26° N latitude and 72 and 80° NE longitude comprising central and peninsular India. African sorghum is grown between 10 and 23° S and 15 and 35° E longitudes. In temperate region, it is grown during summer, whereas in tropics, it can be grown throughout the year. It does not tolerate frost and most sorghum production is concentrated between 40° N and 40° S latitudes. It can adequately recover from drought and can also withstand temporary flooding.

Sorghum is well adopted to semiarid regions with a minimum annual rainfall of 350- 400 mm. It is grown in areas that are too hot and dry for growing maize. It is a drought resistant crop. Sorghum can be grown on wide range of soils. Medium to deep black soils are predominantly suitable for growing sorghum. Rabi sorghum is wholly confined to black soils, while the *kharif* is grown on light soils also. It can be grown with wide range of soil pH from 5.0 to 8.5. It is moderately tolerant to salinity.

Sugar crops: Though sugar crops together occupy slightly >2% of cultivated crop plants in the world. Although sugar beets and sugarcane both produce sugar, but they prefer different climates, with sugarcane favoring year round warmth, while sugar beets favor much cooler condition. The two crops also have different plant physiologies.

Sugarcane is a C_4 crop, while sugar beets are C_3 and in sugarcane the sugar is stored in the stalk, while in beets, it is stored in roots. Sugar beets are cultivated in the temperate latitudes of the N Hemisphere from 40° N to 60° N mostly in Europe and the European part of Russia. While sugarcane is a tropical crop.

5.1. Plant Adaptations

All living organisms on earth are going to react with environmental stimuli and in doing so they show differences in their behavior, morphology, anatomy, physiology and reproduction structures. Such changes which enable the organism to withstand the atrocities of environment and utilize it to their maximum benefit are known as adaptations.

Adaptation is the ability of organism to live and reproduce under congenial or non congenial conditions or Adaptation is the process of adjustment of an organism to survive in its habitat. 'Neger' defined adaptation as that phenomenon by which plants react with the environment through alteration of their inner organization, this reaction leading to the production of more or less expedient characters.

Adaptations of hydrophytes

I. Morphological adaptations:

» Poorly developed root system

» Stem is spongy because of presence of aerenchymatous tissue

» Flowers are produced on long branches above ground surface

» Floating leaves having waxy coating with hydrophobic nature

» Reproduction by vegetative means

» Seeds are light and dispersed by water and sometimes by insects

II. Anatomical adaptations:

» Presence of aerenchymatous tissue in leaves to store air as the plants are under anaerobic conditions

» Extensive air spaces in the stem

» Leaves and roots help in exchange of gases

» Vascular tissues are poorly developed

» In submerged hydrophytes, stomata is absent and in floating hydrophytes, stomata is present on upper side of leaves

III. Physiological adaptations:

» Water and nutrients are absorbed by entire plant body

» Osmotic concentration of cell sap is slightly higher than the water/outside media

» Carbon dioxide released during respiration is stored in air spaces of stem which plant utilizes for photosynthesis. Some air spaces are used for storing oxygen released from photosynthesis.

Adaptations of meso-phytes

I. Morphological adaptations:

» Root system is profusely developed

» Leaves are large in size and horizontally oriented

» Epidermis is thin in meso-phytes

» Colour of the leaves are dark green because of closely packed chlorophyll

II. Anatomical adaptations:

» Plants mainly grow during rainy season

» Mangrove forests have vivi-pary seed germination

» Prop root system is present if few species. *Eg*., maize

» Most of the vegetation along the coastal line and in mangroves has specialized organs for respiration called as "pneumato-phores" (negatively geotropic structures).

III. Physiological adaptations:

» Stomata is present on both sides of leaves

» Conducting tissues are very well developed

» Aerenchymatous tissues are absent

» Some plants have adaptations in relation to soil parameters particularly EC and pH (e.g., osmotic adjustment)

Adaptations of xerophytes

I. Morphological adaptations:

» Root system is extensively developed and sometimes penetrate into very deep layers. *Eg*., Roots of *Prosophis juliflora* can reach a depth of 15-20 m under too dry conditions

» Shoot system is profusely developed

» Leaf size is reduced and some plants do not have leaves at all and some have modified stem. In some plants, leaved are modified (e.g., needles in pine)

II. Anatomical adaptations:

» Have thick cuticle and waxy coating

» Waxy layer on leaves is very common

» Hair like structures are present on leaves to reduce evaporation

» Cell size is reduced

» Powdery like substances are present on leaves to avoid pests and diseases

» Vascular tissues are very well developed

» Stomata are present in cavities and sometimes covered with hairs

III. Physiological adaptations:

» Osmotic concentration of cell sap is always high

» Even at low moisture level, the plants are capable of synthesizing photosynthates

» The plants are elastic and can adjust its life cycle depending on the availability of moisture

» The plants are hardy and drought resistant

» They have short life cycle

» They have less transpiration rate

» Osmotic potential of the plants is very high

Adaptations of plants growing in high altitudes

- They are short growing plants and the plant height is always reduced
- Cell size is also reduced due to low turgidity
- Most of the time cross pollination by insects occurs because of capitulum type of inflorescence
- Flowers of high altitude plants are always colored and relatively bigger and aggregate into clusters and some flowers are even scented to attract insects.
- Interception of solar radiation is very low. Hence, some plants will grow taller.

Adaptations in relation to soil

- Plants on saline soils grow chiefly in rainy season when the soil solution has been diluted and salts move beyond root zone
- Plants are usually shallow rooted
- Surface feeding roots helps in aeration under waterlogged conditions. Pneumato-phores and vivi-pary in mangroves

Adaptations induced by light

- Stem becomes thick and have well developed xylem and phloem
- Plants are profusely branched with smaller internodes
- Cell size and leaf blade is usually reduced. Leaf blade is not flat but oriented other than right angles to the path of incident radiation
- Stomata are smaller and closer together
- Cell wall and cuticle becomes thick
- Chloroplast number becomes less and size also gets reduced.

Chapter - 6

Agro-Climatic Zones of India

Agro-climatic zone is a land unit in terms of major climates suitable for certain range of crops and cultivars (FAO, 1983). The purpose of classifying agro-climatic zones is to developing location specific research and development strategies for increasing agricultural production. In 1988 (Seventh five year plan) the Planning Commission came up with a growth strategy based on a holistic approach of area planning for long-term resource efficiency and sustainability. The objective behind this was that resource based planning becomes feasible once homogeneous regions with respect to natural resource endowments (agro-climatic factors) were delineated and their utilization of available natural resource endowments was related to requirements of output and employment. So, the planning commission delineated India into 15 agro-climatic zones based on homogeneity in rainfall, temperature, topography, water resources, cropping and farming systems (Fig.7). Each zone is divided into 3 to 14 sub zones, totaling 124 sub zones. The classification is largely intended for use in planning and development of agriculture and allied activities. The zones and subz ones (in brackets) are:

1. Western Himalayan Region (11)
2. Eastern Himalayan Region (14)
3. Lower Gangetic Plains Region (4)
4. Middle Gangetic Plains Region (3)
5. Upper Gangetic Plains Region (5)
6. Trans Gangetic Plains Region (6)

7. Eastern Plateau and Hills Region (14)
8. Central Plateau and Hills Region (12)
9. Western Plateau and Hills Region (8)
10. Southern Plateau and Hills Region (13)
11. East Coast Plains and Hills Region (13)
12. West Coast Plains and Hills Region (11)
13. Gujarat Plains and Hills Region (4)
14. Western Dry Region (3)
15. The Islands Region (3)

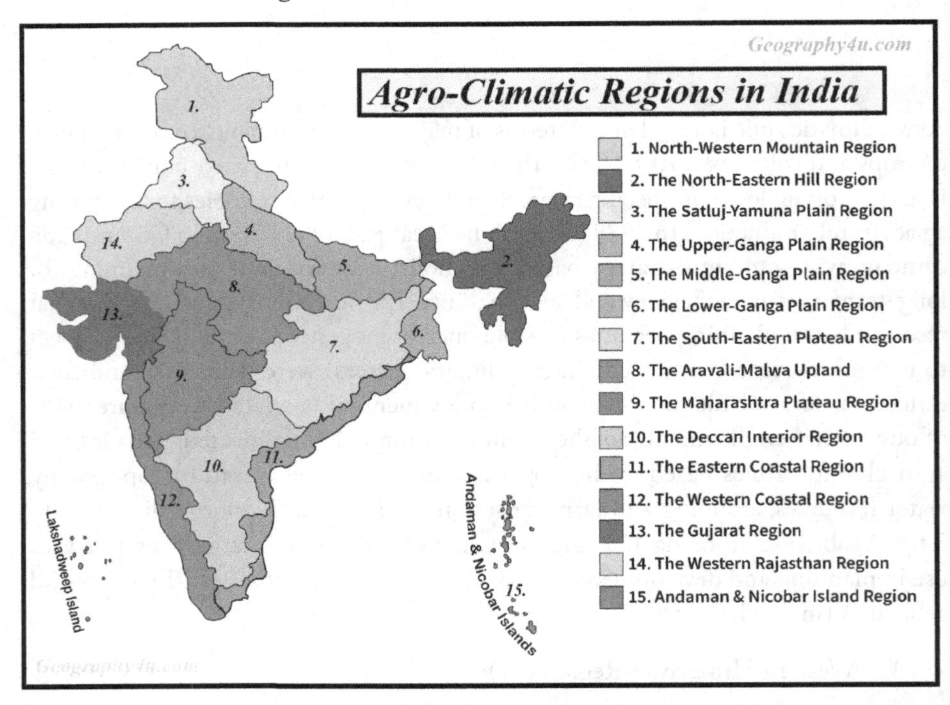

Fig.7. Agro-climatic zones of India

1. Western Himalayan Region: It includes Jammu and Kashmir, Himachal Pradesh and Kumaun-Garhwal areas of Uttaranchal. It shows great variation in relief. Summer season is mild (July average temperature 5 °C-30 °C) but the winter season experiences severe cold conditions (January temperature 0°C to -4 °C). The amount of average annual rainfall is 150 cm. Zonal arrangement in vegetation is found with varying height along the hill slopes. Valleys and duns have thick layers of alluvium while hill slopes have thin brown hilly soils.

The region is gifted with a number of perennial streams due to high rainfall and snow covered mountain peaks of which Ganga, Yamuna, Jhelum, Chenab, Satluj and Beas etc. are worthy of mention. These provide irrigation water to canals and cheap hydel power for agriculture and industries. In recent years the increasing human interference has led to the deterioration of the ecological system. Rapid construction activities in the form of buildings and transport routes and unscrupulous mining have led to the depletion of vegetal cover making the area prone to landslides, pollution, receding glaciers and water scarcity.

Rice is the main crop of this region which is cultivated in terraced fields along the hill slopes. Maize, wheat, potato, barley are other important crops. Temperate fruits like apple and pear are produced in some parts of Jammu and Kashmir and Himachal Pradesh. Similarly tea plantations have started in some areas of Garhwal-Kumaun hills. The horticulture of the region is obsessed with the problems of financial crisis, long gestation period, lack of improved varieties of plants and high post-harvest losses (about 20% in packing, storage, marketing and processing). This needs improvement in old unproductive orchards using the recent technique of grafting, introduction of new fast growing varieties of plants and development of infrastructural facilities. Environmental conditions of this region are well suited for the development of garden and plantation crops. A more rational land use planning is required for the region. Land suitable for agriculture, horticulture, pasture, forestry respectively should be demarcated on the following basis:

a. Agriculture is possible in the lands of upto 30 per cent slopes

b. Land having 30-50 per cent slopes for horticulture/fodder development

c. All lands above 50 per cent slopes under tree cover. Better quality planting material should be made available to fruit growers. The region has favourable climatic conditions for growing temperate vegetables, flowers, and crops like ginger and saffron.

2. Eastern Himalayan Region: The Eastern Himalayan region consists of Sikkim, Darjeeling area (West Bengal), Arunachal Pradesh, Assam hills, Nagaland, Meghalaya, Manipur, Mizoram and Tripura. It is characterized by rugged topography, thick forest cover and sub-humid climate (rainfall over 200 cm; temperature July 25 °C-33 °C, January 11 °C-24 °C). The soil is brownish, thick layered and less fertile. Shifting cultivation (Jhum) is practiced in nearly 1/ 3 of the cultivated area and food crops are raised mainly for sustenance. Rice, potato, maize, tea and fruits (orange, pine-apple, lime, lichi etc.) are the main crops. The ecological system of the region is problematic.

The region needs marked improvement in infrastructural facilities to accelerate the pace of development. Soil degradation should be arrested by controlling deforestation and by terracing in hills. The run-off should be checked and water conserved should be utilized in minor irrigation. Shifting cultivation should be controlled by encouraging permanent settlement. A programme to grow fruits above 30% slopes in the hills needs to be framed. Supporting activities of sericulture, handicrafts poultry, and piggery etc. should be promoted. A long term quality seed production plan should be implemented to assure adequate supply of quality seeds to the farmers

3. Lower Gangetic Plains Region: This region spreads over eastern Bihar, West Bengal and Assam valley. Here average amount of annual rainfall lies between 100 cm-200 cm. Temperature for July month varies from 26 °C-41 °C and for January month 9 °C-24 °C.

The region has adequate storage of ground water with high water table. Wells and canals are the main source of irrigation. The problem of water logging and marshy lands is acute in some parts of the region. Rice is the main crop which at times yields three successive crops (Aman, Aus and Boro) in a year. Jute, maize, potato, and pulses are other important crops. Planning strategies include improvement in rice farming, horticulture (banana, mango and citrus fruits), pisciculture, poultry, livestock, forage production and seed supply. An export processing zone for marine and sea foods needs to be established which should be equipped with modern facilities of freezing, canning, dehydration, and quality control.

Organizational set-up for implementation of ecological plans should include:

a. A committee at the State level to guide and monitor implementation

b. The Zila Parishads at the district level

c. A village level committee under the aegis of the Panchayat backed by trained staff

4. Middle Gangetic Plains Region: It incorporates eastern Uttar Pradesh and Bihar (except Chotanagpur plateau). It is a fertile alluvial plain drained by Ganga River and its tributaries. The average temperature of July month varies from 26 °C- 41 °C and that of January month 9 °C-24 °C. The amount of annual rainfall lies between

100 cm and 200 cm. The region has vast potential of ground water and surface runoff in the form of perennial rivers which is utilized for irrigation through tube wells, canals and wells. Rice, maize, millets in *kharif*, wheat, gram, barley, peas, mustard and potato in *rabi* are important crops.

The agricultural strategy included introduction of modern market oriented agriculture in place of the traditional one. Efforts should be made to improve and stabilize yield of *kharif* paddy which accounts for over 40 per cent of gross cropped area in the region. Similarly increasing maize production, replacing upland paddy and millets with fruits, vegetables and *kharif* pulses (Arahar), encouraging pre-*kharif* paddy (January to June) or maize in low lands, improving low land cultivation through water harvesting structures and alternative farming systems, and utilizing chaur lands for pisciculture are some other suggested measures to boost up agricultural production.

The strategy also includes reclamation of 5.5 lakh ha of usar lands, 25.4 lakh ha of wastelands, and 16.5 lakh ha of fallow lands for agriculture and allied activities (agro-forestry, silviculture, floriculture etc.). Supplementary measures should include input delivery system, demonstration of recommended package of practices, particularly for *kharif* paddy, diversification of crops like vegetables and fruits, and provisions of processing and marketing facilities, poultry, dairying and inland riverine fishery.

5. Upper Gangetic Plains Region: This region encompasses central and western parts of Uttar Pradesh. The climate is sub-humid continental with July month's temperature between 26°-41 °C, January month›s temperature between 7°- 23 °C and average annual rainfall between 75 - 150 cm. The soil is sandy loam. It has 131 per cent irrigation intensity and 144 per cent cropping intensity. Canal, tube well and wells are the main source of irrigation. This is an intensive agricultural region where in wheat, rice, sugarcane, millets, maize, gram, barley, oilseeds, pulses and cotton are the main crops.

Besides modernizing traditional agriculture the region needs special focus on dairy development and horticulture. Strategies should include developing multiple cropping patterns including rice, potato, wheat, moong, sugarcane, sunflower and potato + mustard; improving irrigation system and water management such as lining of canals to check seepage; reclaiming saline/alkaline soils; raising fruit trees on Diara areas; devoting uplands to horticulture; improving cattle breed and bringing more areas to fodder crops.

6. Trans Gangetic Plains Region: The Trans Ganga Plain consists of Punjab, Haryana, Delhi, Chandigarh and Ganganagar district of Rajasthan. The climate has semiarid characteristics with July month's temperature between 26 °C and 42 °C, January temperature ranging from 7 °C to 22 °C and average annual rainfall between 70 cm and 125 cm. Private tube wells and canals provide principal means of irrigation. Important crops include wheat, sugarcane, cotton, rice, gram, maize, millets, pulses and oilseeds etc.

The region has the credit of introducing Green Revolution in the country and has adopted modern methods of farming with greater degree of mechanization. The region is also facing the menacing problem of water logging, salinity, alkalinity, soil erosion and declining water table. The suggested strategies include:

a. Diversion of 5 per cent of rice-wheat area to other crops like maize, pulses, oilseeds and fodder,

b. Development of genotypes of rice, maize and wheat with inbuilt resistance to pests and diseases,

c. Promotion of horticulture besides pulses like tur and peas in upland conditions,

d. Cultivation of vegetables in the vicinity of industrial clusters,

e. Supply of quality seeds of vegetables and planting material for horticulture crops,

f. Development of infrastructure of transit godowns and processing to handle additional fruit and vegetable production,

g. Implementation of policy and programmes to increase productivity of milk and wool, and

h. Development of high quality fodder crops and animal feed by stepping up area under fodder production by 10 percent.

7. Eastern Plateau and Hills Region: It comprises the Chotanagpur plateau Rajmahal hills, Chhattisgarh plains and Dandakaranya. The region enjoys 26 °C-34 °C of temperature in July, 10 °C-27 °C in January and 80 cm-150 cm of annual rainfall. Soils are red and yellow with occasional patches of laterites and alluviums. The region is deficient in water resources due to plateau structure and non-perennial streams. Rainfed agriculture is practised growing crops like rice, millets, maize, oilseeds, ragi, gram and potato.

The region requires planning to maximize use of rainwater, increase in ground water potential, change cropping pattern to achieve a balanced crop production and strengthen input and services delivery systems,

Suggested strategies include:

a. Coverage of large areas with quality seeds of HYV,

b. Cultivation of high value crops of pulses like tur, soyabean, and gram etc. on upland rained areas,

c. Growing crops like urad, castor, and groundnut in kharif and mustard and vegetables in irrigated areas,

d. Improvement of indigenous breeds of cattle and buffaloes,

e. Rehabilitation of 30 per cent of degraded forest lands,

f. Extension of fruit plantations,

g. Renovation including desalting of existing tanks and excavation of new tanks,

h. Reclamation of 95.32 lakh ha of acidic lands through lime treatment,

i. Development of inland fisheries in permanent water bodies, and

j. Adopting integrated watershed development approach to conserve soil and rain water.

8. Central Plateau and Hills Region: This region spreads over Bundelkhand, Baghelkhand, Bhander plateau, Malwa plateau and Vindhyachal hills. The climate is semi-arid in western part to sub-humid in eastern part with temperature in July month 26 °C-40 °C, in January month 7 °C-24 °C and average annual rainfall from 50 cm- 100 cm. Soils are mixed red, yellow and black growing crops like millets, gram, barley, wheat, cotton, sunflower, etc.

The region has dearth of water resources. The suggested measures should include water conservation through water saving devices like sprinklers and drip system; dry farming popularization; dairy development, crop diversification, ground water development, diversion of 16 lakh ha of low value crops to high value crops, reclamation of ravine lands and improvement in indigenous breed cattle.

9. Western Plateau and Hills Region: This region of comprises southern part of Malwa plateau and Deccan plateau (Maharashtra). This is a region of the regard soil with July temperature between 24 °C-41°C, January temperature between 6 °C- 23 °C and average annual rainfall of 25 cm-75 cm. Net sown areas is 65 per cent and forests occupy only 11 per cent. Only 12.4 per cent area is irrigated. Jowar, cotton, sugarcane, rice, bajra, wheat, gram, pulses, potato, groundnut and oilseeds are the principal crops. The area is known for its oranges, grapes and bananas.

Since the region is deficient in water resources attention to be paid for increasing water efficiency by popularizing water saving devices like sprinklers and drip system. The lower value crops of jowar, bajra and rainfed wheat should give way to high value oilseeds like sunflower, safflower and mustard. Five per cent area under rainfed cotton and jowar could be substituted with fruits like ber, pomegranate, mango and guava, appropriate marketing, storage and processing infrastructure to handle added fruit production should be developed. Improvement of milk production of cattle and buffalo through cross-breeding along with poultry development should be taken care of.

10. Southern Plateau and Hills Region: It incorporates southern Maharashtra, Karnataka, western Andhra Pradesh and northern Tamil Nadu. The temperature of July month lies between 26 °C to 42 °C, that of January month between 13 °C-21 °C with annual rainfall between 50 cm-100 cm. The climate is semi-arid with only 50 per cent of area cultivated, 81 per cent of dry land farming, and low cropping intensity of 111 per cent. Low value cereals and minor millets predominate. coffee, tea, cardamom and spices are grown along the hilly slopes of Karnataka plateau.

Suggested strategies include:

a. Diversion of 20 per cent of area under jowar/bajra/ragi to groundnut/ sesamum/sunflower,

b. Cultivation of soybean/ safflower on residual moisture after rice,

c. Development of horticulture crops and agri-horti system,

d. Restoration of tank irrigation by taking appropriate measures of deepening, embanking etc.

e. Use of water saving devices like sprinklers and drip system,

f. Development of location specific dry land farming technology, viz., early fertilizer responsive varieties of jowar and rice, and

g. Implementation of programmes for poultry farming and dairy development.

11. East Coast Plains and Hills Region: This region includes the Coromandel and Northern Circar coasts. Here climate is sub-humid maritime with May and January temperatures ranging from 26 °C-32 °C and 20 °C-29 °C, respectively and annual rainfall of 75 cm-150 cm. The soils are alluvial, loam and clay facing the menacing problem of alkalinity. The region accounts for 20.33 per cent of rice and 17.05 per cent of groundnut production of the country. Main crops include rice, jute, tobacco, sugarcane, maize, millets, groundnut and oilseeds.

Main agricultural strategies include improvement in the cultivation of spices (pepper and cardamom) and development of fisheries. These involve increasing cropping intensity using water-efficient crops on residual moisture, discouraging growing of rice on marginal lands and bring such lands under alternate crops like oilseeds and pulses; diversifying cropping and avoiding mono-cropping; developing horticulture in upland areas; increasing marine, brackish water and inland fisheries; upgrading genetic potential of existing animals; improving tank irrigation by desalting and embanking existing tanks and improving field channels and structures; and providing solutions to the problems of water logging and alkalinity.

12. West Coast Plains and Ghats Region: This region extends over the Malabar and Konkan coasts and the Sahyadris and is covered by laterite and coastal alluvials. This is a humid region with annual rainfall above 200 cm and average temperatures of 26 °C-32 °C in July and 19 °C-28 °C in January. Rice, coconut, oilseeds, sugarcane, millets, pulses and cotton are the main crops. The region is also famous for plantation crops and spices which are raised along the hill slopes of the Ghats.

The strategies include development of high value crops and fisheries, protection of land from salinity and provision for drainage improvement. A programme to sink about 5 lakh dug wells and shallow wells should be undertaken on priority basis. The approach of homestead system of reclaiming and using Khar lands be accepted, planned and implemented. Low productivity areas of paddy and millets under rained conditions should be diverted to horticultural crops like mango, banana, coconut etc. to cover about 0.5 lakh ha. Appropriate infrastructure for fruit marketing and processing should be developed. Increasing mechanized fishing boats for deep sea fishing, strengthening of cold storage, processing, and transport infrastructure in corporate/cooperative sector, and providing incentives for prawn culture in brackish water should be taken up.

13. Gujarat Plains and Hills Region: This region includes Kathiawar and fertile valleys of Mahi and Sabarmati rivers. It is an arid and semi-arid region with average annual rainfall between 50 cm-100 cm, and monthly temperature between 26 °C-42 °C in July and 13 °C-29 °C in January. Soils are regur in the plateau region, alluvium in the coastal plains, and red and yellow soils in Jamnagar. Groundnut, cotton, rice, millets, oilseeds, wheat and tobacco are the main crops. It is an important oilseed producing region.

The major thrust of development in the region should be on canal and ground water management, rain water harvesting and management, dry land farming, agro-forestry development, wasteland development and developing marine fishing and brackish/back-water aquaculture in coastal zones and river deltas.

14. Western Dry Region: It comprises western Rajasthan west of the Aravallis. It is characterized by hot sandy desert, erratic rainfall (annual average less than 25 cm), high evaporation, contrasting temperature (June 28 °C- 45 °C, and January 5 °C-22 °C), absence of perennial rivers and scanty vegetation. Ground water is very deep and often brackish. Famine and droughts are common features. Land-man ratio is high. Forest area is only 1.2 per cent. Land under pastures is also low (4.3 per cent). Cultivable waste and fallow lands account for nearly 42 per cent of the geographical area. Net irrigated area is only 6.3 per cent of net sown area which is 44.4 per cent of the geographical area. Bajra, jowar, and moth are main crops of *kharif* and wheat and gram in rabi. Livestock contributes greatly in desert ecology.

Agricultural development strategies include completion of on-going irrigation projects; making research on the use of saline water; popularizing bajra varieties giving higher biomass; promoting use of fertilizers; making improved seeds available to farmers, constructing moisture (rain water) conservation structures; increasing yield level of fruits like date palm, water-melon and guava; adopting high quality germplasm in cattle to improve their breed; and adopting silvi-pastoral system over wastelands through aerial and manual seeding.

15. The Islands Region: The island region includes Andaman-Nicobar and Lakshadweep which have typically equatorial climate (annual rainfall less than 300 cm the mean July and January temperature of Port Blair being 30 °C and 25 °C respectively). The soils vary from sandy along the cost to clayey loam in valleys and lower slopes. The main crops are rice, maize, millets, pulses, areca nut, turmeric and cassava. Nearly half of the area is under coconut. The area is covered with thick forests and agriculture is a backward stage.

The main thrust in development should be on crop improvement, water management and fisheries. Improved varity of paddy seeds should be popularized so as to enable farmers to take two crops of rice in place of one. For fisheries development multi-purpose fishing vessels for deep sea fishing should be introduced, suitable infrastructure for storage and processing of fish should be built up, and brackish water prawn culture should be promoted in the coastal area.

Agro-ecological Zones of India

Identification of homogeneous agro-ecological zones is necessary for sustainable utilization of land, water and other natural resources for development through transfer of suitable technology. Development planning based on characteristics of agro-ecological zones is much relevant to achieve higher production. The National Bureau of Soil Survey and Land Use Planning (NBSS&LUP) of the ICAR has divided India into 20 agro ecological regions in 1999 on the basis of physiography (topography and drainage), soil, bioclimate and length of crop growing period (Fig.8).

Agro-ecological zone is a land unit carved out of climatic zone, correlated with landforms, climate and the length of growing period (LGP). LGP refers to the number of days available for crop growth with suitable conditions.

1. Cold Arid Eco-region with Shallow Skeletal Soils

2. Hot Arid Eco-region with Desert and Saline Soils

3. Hot Arid Eco-region with Red and Black Soils

4. Hot Semi-Arid Eco-region with Alluvium Derived soils

5. Hot Semi Arid Eco-region with Medium and Deep Black Soils

6. Hot Semi-Arid Eco-region with Shallow and Medium (Dominant) Black Soils

7. Hot Semi Arid Eco-region with Red and Black Soils

8. Hot Semi-Arid Eco-region with Red Loamy Soils

9. Hot subbumid (Dry) Eco-region with Alluvium- Derived Soils

10. Hot Subhumid Eco-region with Red and Black Soils

11. Hot Subhumid Eco-region with Red and Yellow Soils

12. Hot Subhumid Eco-region with Red and Lateritic Soils

13. Hot Suhhumid (Moist) Eco-region with Alluvium-derived Soils

14. Warm Subhumid to Humid with Inclusion of Perhumid Eco-region with Brown Forest and Podzolic Soils

15. Hot Subhumid (moist) to Humid (inclusion) of perhumid Eco-region with alluvium-derived Soils

16. Warm Perhumid Eco-region with Brown and Red Hill Soils

17. Warm Perhumid Eco-region with Red and Lateritc Soils

18. Hot Subhumid to Semi-arid Eco-region with Coastal Alluvium-derived Soils

19. Hot Humid Pemhumid Eco-region with Red, Lateritic and Alluvium-derived Soils

20. Hot Humid/Perhumid Island Eco-region with Red loamy and Sandy Soils

Fig.8: Major agro-ecological regions of India

The agro-ecological regions are derived from ecosystems. Major advantage of length of crop growing season (LCGS) or length of growing period (LGP) is that it is the direct indicative of given landform rather than total rainfall. For example, both Ratnagiri in western Maharashtra and Nagpur in eastern Maharashtra have LGP 180 to 210+ days but the total annual rainfall of Ratnagiri is more than 2,000 mm where as that of Nagpur is only 1,100 mm. Therefore, crop planning has to be based on LGP rather than total rainfall.

Characteristic futures of the 20 agro-ecological regions (AERS) are summarized.

1. Arid ecosystem

1. Western Himalayas, cold arid eco-region, with shallow skeletal soils and length of growing period (GP) < 90 days.
2. Western Plain, Kachchh and part of Kathiawar Peninsula, hot arid eco-region, with desert and saline soils and GP < 90 days.
3. Deccan Plateau, hot arid eco-region, with red and black soils and GP < 90 days.

2. Semi-arid ecosystem

4. Northern Plain and Central Highlands including Aravallis, hot semiarid ecoregion, with alluvium derived soils and GP 90-150 days.
5. Central (Malwa) Highlands, Gujarat Plains and Kathiawar Peninsula, hot semiarid eco-region, with medium and deep black soils and GP 90-150 days.
6. Deccan Plateau, hot semiarid eco-region with shallow and medium (with inclusion of deep) black soils and GP 90-150 days.
7. Deccan (Telangana) Plateau and Eastern Ghats, hot semiarid eco-region, with red and black soils and GP 90-150 days.
8. Eastern Ghats, TN uplands and Deccan (Karnataka) Plateau, hot semiarid eco-region with red loamy soils and GP 90-150 days.

3. Sub-humid ecosystem

9. Northern Plain, hot sub-humid (dry) eco-region, with alluvium-derived soils and GP 150-180 days.
10. Central Highlands (Malwa, Bundelkhand & Satputra), hot sub-humid eco-region, with black and red soils and GP 150-180 (to 210) days.
11. Eastern Plateau (Chhatisgarh), hot sub-hmid eco-region, with red and yellow soils and GP 150-180 days.
12. Eastern (Chhotanagpur) Plateau and Eastern Ghats, hot sub-humid eco-region, with red and lateritic soils and GP 150-180 (to 210) days.
13. Eastern Plain, hot sub-humid (moist) eco-region, with alluvium-derived soils and GP 180-210 days.

14. Western Himalayas, warm sub-humid to humid with inclusion of perhumid eco-region with brown forest and podzolic soils and GP 180-210+ days.

4. Humid-perhumid ecosystem

15. Bengal and Assam Plain, hot sub-humid (moist) to humid (inclusion of perhumid) eco-region, with alluvium-derived soils and GP 210+ days.
16. Eastern Himalayas, warm perhumid eco-region, with brown and red hill soils and GP 210+ days.
17. North-eastern Hills (Purvachal), warm perhumid eco-region, with red and lateritic soils and GP 210+ days.

5. Coastal ecosystem

18. Eastern Coastal Plain, hot sub humid to semiarid eco-region, with coastal alluvium-derived soils and GP 90-210+ days.
19. Western Ghats and Coastal Plain, hot humid perhumid eco-region, with red, lateritic and alluvium-derived soils and GP 219 + days.

6. Island ecosystem

20. Islands of Andaman-Nicobar and Lakshadweep hot humid to perhumid island eco-region, with red loamy and sandy soils and GP 210+ days.

Chapter - 7

Ecosystem

We know that earth is perhaps the only planet in the solar system that supports life. The portion of the earth which sustains life is called biosphere. Biosphere is very huge and cannot be studied as a single entity. It is divided into many distinct functional units called ecosystems. The term 'ecosystem' was coined by A.G. Tansley in 1935. The term `eco' refers to a part of the world and `system' refers to the coordinating units. Ecosystem is a result of interaction of all living organisms and non living factors associated with the environment.

Definitions of ecosystems may have differences among authors but generally all have three common properties that include the presence of (1) biotic and (2) abiotic components and their (3) interactions. Although all definitions of ecosystem contain the three aforementioned components, the differences in definitions are the ways of expressing. It is instructive to consider three definitions of ecosystem that reflect changes in the concept since its first use. A. G. Tansley coined the term "ecosystem" as part of a debate over the nature of biological communities.

Tansley defined ecosystem as "a particular category of physical systems, consisting of organisms and inorganic components in a relatively stable equilibrium, open and of various sizes and kinds". According to Tansley, the ecosystem is comprised of two major parts *viz.*, biome and habitat and thus all parts of such an ecosystem-organic and inorganic, biome and habitat may be regarded as interacting factors which, in a mature ecosystem, are in approximate equilibrium, it is through their interactions that the whole system is maintained.

According to E. P. Odum "any unit that includes all the organisms (*i.e.*, the "community") in a given area interacting with the physical environment so that a flow of energy leads to clearly defined trophic structure, biotic diversity and material cycles (*i.e.*, exchange of materials between living and nonliving parts) within the system in an ecological system or ecosystem. F. R. Fosberg (1963) has defined ecosystem as a functioning and interacting system composed of one or more living organisms and their effective environment, both physical and biological. According to R. L. Lindeman (1942) **the term ecosystem applies to any system composed of physical-chemical-biological processes, within a space-time unit of any magnitude. According to** A. N. Strahler and A. H. Strahler (1976), the total assemblage of components interacting with group of organisms is known as ecological system or more simply, an ecosystem. An ecosystem, according to Smith (1970) is "a functional unit with recognizable boundaries and an internal homogeneity". Ellenberg (1973) describes an ecosystem as "an interacting system between organisms and their inorganic environment which is open but has a certain degree of ability of self regulation." According to Woodbury (1954), ecosystem is a complex unit in which habitat, plants and animal are considered as one interesting unit, the materials and energy of one passing in and out of the others.

There are many other parallel terms or synonyms for the ecosystem which have been proposed by various ecologists *e.g.*, biocoenonsis (Karl Mobius, 1877), microcosm (S. A. Forbes, 1887), holocoen (Friederichs, 1930), biosystem (Thienemann, 1939), geobiocoenosis (Sukhachev, 1944), bioenert body (Vernadsky, 1944) and ecosom, etc.

In the subject of ecology, the term ecosystem refers to the environment of life. An ecosystem includes all the organisms and the nonliving environment that are found in a particular place. In an ecosystem, there are various levels of organization. The simplest level of organization in ecosystem is that of the organism. An organism refers to a particular species in an ecosystem, say cat, dog, plant etc. A population includes all the members of the same organism that live in one place at one time. All the different populations that live in a particular area make up a community. The physical location of a community is called the habitat. Ecosystem is in turn a level of organization and has one higher level of organization called biosphere. The diversity of an ecosystem is a measure of the number of different species present and how common each species is. Ecosystems are very complex and they can contain hundreds or even thousands of interacting species. Each organism or species in the community has a role or profession in that community and in ecology this is the organism's niche. An ecosystem may be conceived and studied in the habitat of various sizes *e.g.*, one square meter of grassland, a pool, a large lake, a large tact of forest, balanced aquarium, a certain area of river and ocean. All the ecosystems of the earth are connected to one another *e.g.*, river ecosystem is connected with the ecosystem of oceans and a small ecosystem of dead logs in a part of large ecosystem of forest. A complete self sufficient ecosystem is rarely found in nature but situations approaching self sufficiency may occur.

Approachs to ecosystem

With an ecosystem comprising of large number of species, it would seem and is impractical to study the interaction of each organism with another. It is impossible to approach an ecosystem by studying the individual organism–environment relationship. Therefore we study an ecosystem following a wholesome approach.

We study the ecosystems by studying the two aspects (attributes) of an ecosystem. They are:

(1) Structure or architectural process

(2) Function or working process

The structure is related to the species diversity. The more complex is the structure, the greater is the diversity of the species in the ecosystem. The function of the ecosystem is related to the flow of energy and cycling of materials through structural components of ecosystem.

Structure of an ecosystem

By architecture or structure of an ecosystem, we mean

» The composition of biological community including species, numbers, biomass, life history and distribution in space, etc.

» The quantity and distribution of non living materials like nutrients, water etc.

» The conditions of existence such as temperature, light etc.

An ecosystem possesses both biotic and abiotic factors. The nonliving factors, called abiotic factors are physical and chemical characteristics of the environment. The living components of the environment are called biotic factors.

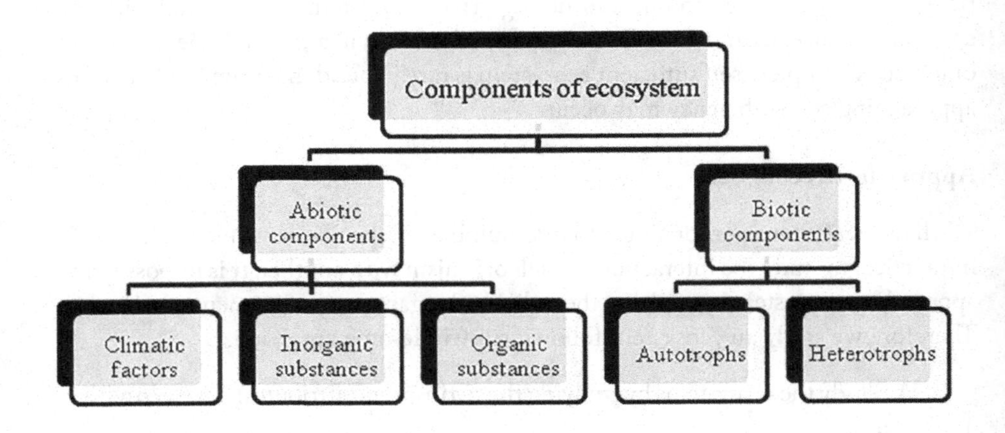

A. Abiotic components: Abiotic components of the ecosystem comprise three sorts of components.

1. **Climatic factors:** These are the physical factors of the given region such as air, water, soil, temperature, light (i.e., its duration and intensity), rainfall, moisture, relative humidity, pH *etc...*

2. **Inorganic substances:** The amount of inorganic substances present at any given time in an ecosystem is designated as the standing state or standing quality. It includes rock, soil, water, oxygen, CO_2, nitrogen (N), phosphorus (P), sulphur (S) and other minerals.

3. **Organic substances:** They are the building blocks of living system and therefore, make a link between the biotic and abiotic components. They include proteins, carbohydrates, lipids and humic substances.

B. Biotic components: In the trophic structure of any ecosystem, living organisms are distinguished on the basis of their nutritional relationships, which are discussed as follows:

1. **Autotrophs** (*auto* = self; *trough* = nourishing): Autotrophic component of ecosystem includes the producers or energy transducers which convert solar energy into chemical energy (that becomes locked in complex organic substances such as carbohydrates, lipids, proteins, *etc.*) with the help of simple inorganic substances such as water and carbon dioxide and organic substances such as enzymes. Autotrophs fall into following two groups:

i. **Photo-autotrophs** which contain green photosynthetic pigment chlorophyll to transduct the solar or light energy of sun e.g., trees, grasses, algae, other tiny phytoplanktons, photosynthetic bacteria and cyano-bacteria (blue green algae).

ii. **Chemo-autotrophs** which use energy generated in oxidation-reduction process, but their significance in the ecosystem as producers is minimal. *E.g.*, Micro-organisms such as *Beggiatoa*, sulphur bacteria, *etc.*

1. **Heterotrophs** (*hetero* = other; *trophic* = nourishing): Heterotrophic organisms predominate the activities of utilization, rearrangement and decomposition of complex organic materials. Heterotrophic organisms are also called as consumers, as they consume the matter built up by the producers (autotrophs). The consumers are of following two main types:

i. **Macro-consumers**: These are also called phagotrophs (*phago* = to eat) and include mainly animals which ingest other organisms or chunks of organic matter. Depending on the food habits, consumers may either be herbivores (plant eaters) or carnivores (flesh eaters). Herbivores live on living plants and are also known as primary consumers, e.g., insects, zooplankton, animals such as rat, deer, cattle, elephant, etc. Secondary and tertiary consumers in the ecosystem are carnivores or omnivores, e.g., insects such as preying mantis, dragon flies; spiders and large animals such as tiger, lion, leopard, wolf, etc. Secondary consumers are the carnivores which feed on primary consumers (*e.g.*, snake feeding on rat) and tertiary consumers feed on secondary consumers (*e.g.*, eagle feeding on snake).

Ticks, mites, leeches and blood sucking insects (mosquito, bed bug) are dependent on herbivores, carnivores and omnivores.

ii. **Micro-consumers:** These are also called decomposers, reducers, saprotrophs/saprophytes (*sapro* = decompose), osmotrophs and scavengers. Wiegert and Owen (1971) have coined the term, biophages for heterotrophic decomposers which feed on the dead organic matter. Micro-consumers include microorganisms such as bacteria, actinomycetes and fungi. Micro-consumers breakdown complex organic compounds of dead or living protoplasm, absorb some of the decomposition or breakdown products and release inorganic nutrients in the environment, making them available again to autotrophs or producers. Some invertebrate animals such as protozoa, oligochaeta such as earthworms, *etc.*, use the dead organic matter for their food, as they have the essential enzymes and hence, can be classified as decomposer organisms. Some ecologists believe that microorganisms are primary decomposers, while invertebrates are secondary decomposers.

The disintegrating dead organic matter is also known as organic detritus (Latin word *deterere* means to wear away). By the action of detritivores, the disintegrating detritus result into particulate organic matter and dissolved organic matter which play important role in the maintenance of the edaphic environment.

Example of a pond ecosystem

A pond as a whole serves as a good example of aquatic and fresh water ecosystem (Fig.9). In fact, it represents a self sufficient and self regulating system. It has the following components:

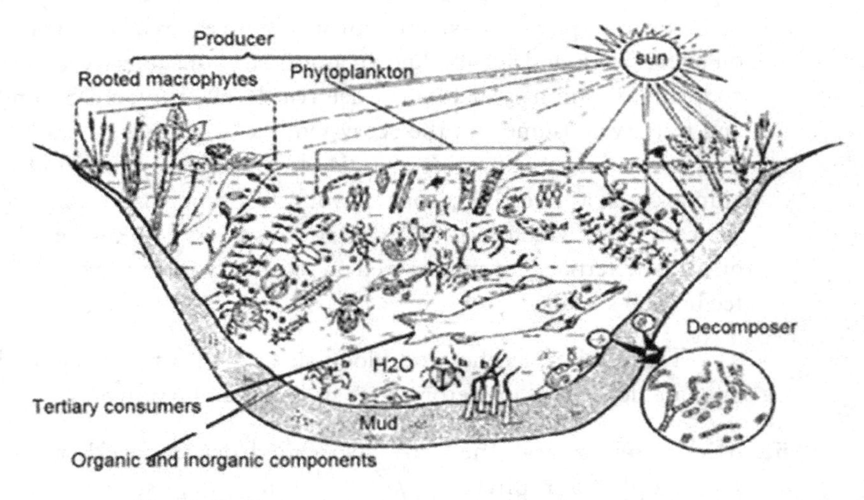

Fig.9: Different components in pond ecosystem

A. Abiotic component:

1. **Climatic/physical factors:** Solar radiation provides energy which is major climatic factor that controls the entire system. The penetration of light depends on transparency of water, amount of dissolved or suspended particles in water and the number of plankton. On the basis of extent of penetration of light a

pond can be divided into euphotic, mesophotic and aphotic zones. Plenty of light is available to plants and animals in euphotic zone. No light is available in the aphotic zone.

2. **Inorganic substances**: These are water, carbon, nitrogen, phosphorus, calcium and a few other elements like sulphur depending on the location of the pond. The inorganic substances like O_2 and CO_2 are in dissolved state in water. All plants and animals depend on water for their food and exchange of gases, nitrogen, phosphorus, sulphur and other inorganic salts are held in reserve in bottom sediment and inside the living organisms. A very small fraction may be in the dissolved state.

3. **Organic compounds**: The commonly found organic matter in the pond are amino acids and humic acids and the breakdown products of dead animals and plants. They are partly dissolved in water and partly suspended in water.

B. Biotic components: It includes various organisms which are classified into following types:

1. **Producers:** These are photo-autotrophic green plants and photosynthetic bacteria. The producers fix radiant energy of sun and with the help of minerals derived from water and mud, they manufacture complex organic substances such as carbohydrates, proteins and lipids. Producers of pond are of following types:

    i. **Macro-phytes**: These include mainly the rooted large sized plants which comprise three types of hydrophytes partly or completely submerged, floating and emergent aquatic plants. The common submerged plants are species of *Potamogeton, Chara, Hydrilla, Vallisneria, Utricularia* and emergent aquatic species are *Typha, Sagittaria, Nymphaea, Nelumbo, etc.* Besides these plants, some free floating forms also occur in the pond ecosystem, *e.g., Azolla, Salvinia, Wolffia, Eichhornia, Spirodella, Lemna,* etc.

    ii. **Phyto-planktons:** These are microscopic, floating or suspended lower plants (algae) that are distributed throughout the water, but mainly in the euphotic zone. Most of them are filamentous algae such as *Spirogyra, Ulothrix, Zygnema, Cladophora* and *Oedogonium*. There also occur some chlorococcales (*e.g., Chlorella*), *Closterium, Cosmarium, Eudorina, Pandorina, Pediastrum, Scendesmus, Volvox,* Diatoms, *Anabaena, Gloeotrichia, Microcystis, Oscillatoria, Chlamydomonas, Spriulina, etc.*, and some flagellates.

2. **Macro-consumers:** They are phagotrophic heterotrophs which depend for their nutrition on the organic food manufactured by producers, the green plants. Macro-consumers are of following three types :

   i. **Herbivores (Primary consumers):** These animals feed directly on living plants (producers) or plant remains. They may be large or minute in size and are of following two types:

      1. **Benthos** which are the bottom dwelling forms such as fish, insect larvae, beetles, mites, molluscs (*e.g., Pila, Planorbis, Unio, Lamellidens, etc.*), crustaceans, *etc.*

      2. **Zooplanktons** which feed chiefly on phyto-planktons and are chiefly the rotifers such as *Brachionus, Asplanchna, Lecane, etc.*, although some protozoans as *Euglena, Coleps, Dileptus, etc.*, and crustaceans such as *Cyclops, Stenocypris*, etc., are also present in the pond.

Besides these small sized herbivores, some mammals such as cow, buffaloes, *etc.*, also visit the pond casually and feed on marginal rooted macrophytes. Some birds also regularly visit the pone feed on some hydrophytes.

   ii. **Secondary consumers** (Carnivore order-1): These carnivores feed on the herbivores and include chiefly insects, fish and amphibians (frog). Most insects are water beetles which feed zooplanktons; some insects are the nymphs of dragonflies which feed upon aquatic insects.

   iii. **Tertiary consumers** (Carnivore order-2): These are some large fish as game fish that feed on the smaller fish and thus, become the tertiary (top) consumers.

3. **Decomposers:** They are also called micro-consumers, since they absorb only a fraction of the decomposed organic matter. They bring about the decomposition of dead organic matter of both producers (plants) as well as macro-consumers (animals) to simple forms. Decomposers help in returning of mineral elements again to medium of the pond and in running biogeochemical cycles. Decomposers of pond ecosystem include chiefly bacteria, actinomycetes and fungi. Among fungi, species of *Aspergillus, Cephalosporium, Cladosporium, Pythium, Rhizopus, Penicillium, Thielavia, Alternaria, Trichoderms, Circinella, Fusarium, Curvularis, Paecilomyces, Saprolegnia, etc.*, are most common decomposers in water and mud of the pond.

7.1. Functions of an ecosystem

Ecosystems are complex and dynamic. They perform certain functions and they are:

i. Productivity in an ecosystem
ii. Energy flow
iii. Nutrient cycling (biogeochemical cycles)
iv. Ecological succession
v. Homeostasis

I. Productivity in an ecosystem: The productivity of an ecosystem refers to the rate of production, *i.e.*, the amount of organic matter accumulated in any unit time. It is of following types

1. **Primary productivity**. It is defined as the rate at which radiant energy is stored by photosynthetic and chemosynthetic activity of producers. Primary productivity is of following types:

   i. **Gross primary productivity.** It refers to the total rate of photosynthesis including the organic matter used up in respiration during the measurement period. GPP depends on the chlorophyll content. The rate of primary productivity is estimated in terms of either chlorophyll content as chl/g dry weight/unit area or photosynthetic number, *i.e.*, amount of CO_2 fixed/g chl/hour.

   ii. **Net primary productivity**. It is the rate of storage of organic matter in plant tissues in excess of the respiratory utilization by plants during the measurement period.

2. **Secondary productivity**. It is the rate of energy storage at consumer's levels herbivores, carnivores and decomposers. Consumers tend to utilize already produced food materials in their respiration and also convert the food matter to different tissues by an overall process. So, secondary productivity is not divided into 'gross' and 'net' amounts. Due to this fact some ecologists prefer to use the term assimilation rather than production at this level - the consumers level. Secondary productivity, in fact, remains mobile (*i.e.*, keeps on moving from one organism to another) and does not live *in situ* like the primary productivity.

3. **Net productivity**. It is the rate of storage of organic matter not used by the

hetero-trophs or consumers, *i.e.*, equivalent to net primary production minus consumption by the hetero-trophs during the unit period as a season or year, etc.

II. Energy flow in an ecosystem:

Flow of energy in an ecosystem takes place through the food chain and it is this energy flow which keeps the ecosystem going (Fig.10). The most important feature of this energy flow is that it is unidirectional or one-way flow. Unlike the nutrients, (like carbon, nitrogen, phosphorus etc.) energy is not reused in the food chain. Also, the flow of energy follows the two laws of Thermodynamics:

I law of thermodynamics states that energy can neither be created nor be destroyed but it can be transferred from one form to another. The solar energy captured by the green plants (producers) gets converted into biochemical energy of plants and later into that of consumers.

II law of Thermodynamics states that energy dissipates as it is used or in other words, it gets converted from a more concentrated to dispersed form. As energy flows through the food chain, there occurs dissipation of energy at every trophic level.

The flow of energy in an ecosystem is always linear or one way. The quantity of energy flowing through the successive trophic levels decreases as shown by the reduced sizes of boxes in fig. below. At every step in a food chain or web the energy received by the organism is used to sustain itself and the left over is passed on to the next trophic level.

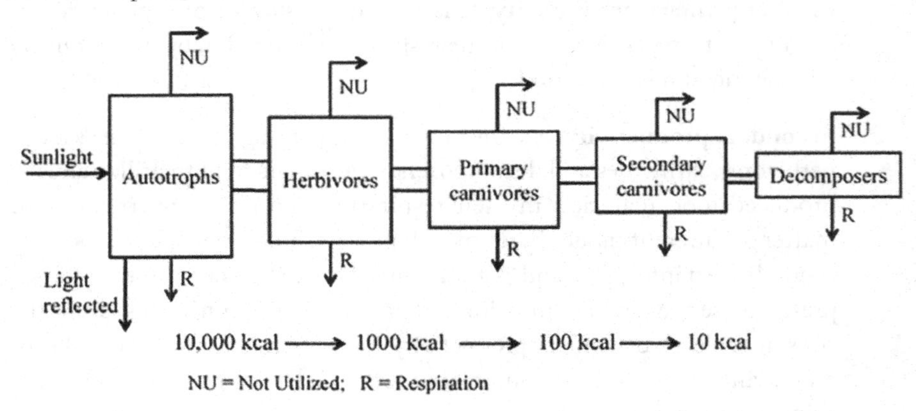

Fig.9: Model of energy flow through an ecosystem. Boxes indicate the standing crop biomass and pipes indicate the energy flowing

7.2. Food Chain

The transfer of food energy from the producers, through a series of organisms (herbivores to carnivores to decomposers) with repeated eating and being eaten, is known as a food chain. Producers utilize the radiant energy of sun which is transformed to chemical form, ATP during photosynthesis. Thus green plants in any food chain occupy the first trophic level and are called the primary producers. The energy as stored in food matter manufactured by green plants is then utilized by the plant eaters *i.e.*, the herbivores, which constitute the second trophic level (primary consumers level) are called the primary consumers (herbivores). Herbivores in turn are eaten by the carnivores, which constitute the third trophic level (secondary consumers level) are called the secondary consumers (carnivores). These in turn may be eaten still by other carnivores at quaternary trophic level or tertiary consumers level and they are referred as tertiary consumers (carnivores). Some organisms are omnivores eating both the primary producers and consumers.

In nature, we come across two major types of food chains:

1. **Grazing food chain:** This type of food chain starts from the living green plants and goes to grazing herbivores and on to carnivores. Ecosystems with such type of food chain are directly dependent on the influx of solar radiation. This type of chain thus depends on autotrophic energy capture and the movement of this captured energy to herbivores. Most of the ecosystems in nature follow this type of food chain. From energy standpoint, these chains are very important. The phytoplanktons – zooplanktons – fish sequence or the grasses – rabbit – fox sequence are the examples of grazing food chain.

2. **Detritus food chain:** This type of food chain goes from dead organic matter into micro-organisms and then to organisms feeding on detritus (detritivores) and their predators. Such ecosystems are thus less dependent on direct solar energy. These depend chiefly on the influx of organic matter produced in another system. For example, such type of food chain operates in the decomposing accumulated litter in a temperate forest. A good example of a detritus food chain is based on mangrove leaves described by Hcald (1969) and W.E. Odum (1970). In the brackish zone of Southern Florida, leaves of the red mangrove, *Rhizopho mangle* fall into the warm, shallow waters. Only 5% of the leaf material was removed by grazing insects before leaf fall. The fallen fragments (acted on by such saprotrophs as fungi, bacteria, protozoa *etc*, and colonized mainly by phytoplanktonic and benthic algae) are eaten and re-eaten (coprophagy) by a key group of small animals. These animals include crabs, copepods, insect larvae,

grass shrimps, mysids, nematodes, amphipods, bivalve mollusks, *etc.* All these animals are detritus consumers. These detritivores are the key group of small animals, comprising only a few species but very large number of individuals. They ingest large amounts of the vascular plant detritus. These animals are in turn eaten by some minnows and small game fish *etc. i.e.*, the small carnivores, which in turn serve as the main food for larger game fish and fish eating birds which are the large (top) carnivores. The mangroves considered generally as of less economic value make a substantial contribution to the food chain that supports the fisheries, an important economy in that region. Similarly, detritus from sea grasses, salt marsh grasses and sea weeds support fisheries in many estuarine areas.

Significance of studying food chains

1. It helps in understanding the feeding relations and interactions among different organisms of an ecosystem.
2. It explains the flow of energy and circulation of materials in ecosystems.
3. It helps in understanding the concept of bio-magnification in ecosystems.

Food web: However, food chains in natural conditions never operate as isolated sequences, but are interconnected with each other forming some sort of interlocking pattern, which is referred to as a food web. Under natural conditions the linear arrangement of food chains, hardly occurs and these remain indeed interconnect with each other through different types of organisms at different trophic levels. For example, in grazing food chain of grassland, in the absence of rabbit, grass may also be eaten by mouse. The mouse in turn may be eaten directly by hawk or by snake first which is then eaten by hawk. Thus, in nature there are found alternatives which all together constitute some sort of interlocking pattern *i.e.*, the food web.

In such a food web in grassland, as shown in figure 11, there may be seen as many as five linear food chains, which in sequences are:

1. Grass → Grasshopper → Hawk
1. Grass → Grasshopper → Lizard→ Hawk
1. Grass → Rabbit → Hawk (or vulture or fox or even man, if present)
1. Grass → Mouse → Hawk
1. Grass → Mouse → Snake → Hawk

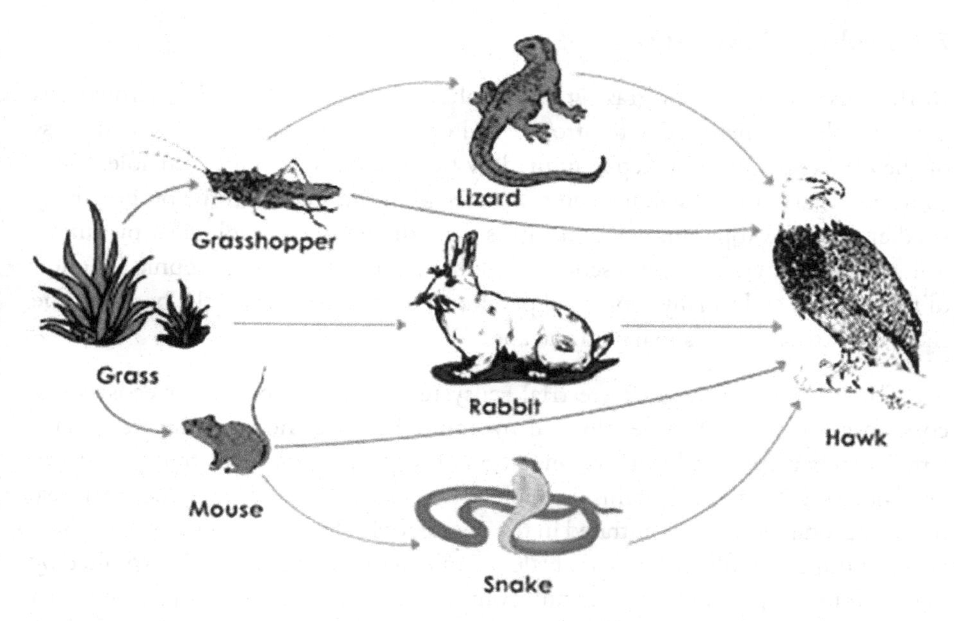

Fig. 11: Diagrammatic sketch showing a food web in a grassland ecosystem

Differences between food chain and food web

Food chain	Food web
Flow of energy through a single straight pathway from one trophic level to other trophic level is called food chain	The interconnected, involving numerous food chains through which the energy flows in an ecosystem is called food web
It consists of only one straight chain	It contains many interconnected food chains
Instability increases due to increase in number of organisms	Stability increases due to presence of complex food chains
Even if one group of organism is disturbed, the whole food will be affected	Food web doesn't get affected much, even if one group of organism is removed
It has 4-6 trophic levels	It has many trophic levels with different populations of species
It is of two types: grazing food chain and detritus food chain	No further classification

7.3. Ecological pyramids

In the successive steps of grazing chain, photosynthetic autotrophs, herbivorous heterotrophs, carnivorous heterotrophs and decomposers - the number and mass of the organisms in each step is limited by the amount of energy available. Since some energy is lost as heat, in each transformation the steps become progressively smaller near the top. This relationship is sometimes called "ecological pyramid". The ecological pyramids represent the trophic structure and also trophic function of the ecosystem. In many ecological pyramids, the producer form the base and the successive trophic levels make up the apex.

Thus, communities of terrestrial ecosystems and shallow water ecosystems contain gradually sloping ecological pyramids because these producers remain large and characterized by an accumulation of organic matter. This trend, however, does not hold for all ecosystems. In such aquatic ecosystems as lakes and open sea, primary production is concentrated in the microscopic algae. These algae have a short cycle, multiply rapidly, accumulate little organic matter and are heavily exploited by herbivorous zooplankton. At any one point in time the standing crop is low. As a result, the pyramid of biomass for these aquatic ecosystems is inverted *i.e.*, the base is much smaller than the structure it supports.

Types of ecological pyramids

The ecological pyramids may be of following three kinds:

i. **Pyramid of number:** It depicts the number of individual organisms at different trophic levels of food chain. This pyramid was advanced by Charles Elton (1927), who pointed out the great difference in the number of the organisms involved in each step of the food chain. The animals at the lower end (base of pyramid) of the chain are the most abundant. Successive links of carnivores decrease rapidly in number until there are very few carnivores at the top. The pyramid of number ignores the biomass of organisms and it also does not indicate the energy transferred or use of energy by the groups involved. The pyramid of biomass can also be upright or inverted depending on the type of ecosystem. Grassland and pond ecosystem show an upright pyramid of numbers.

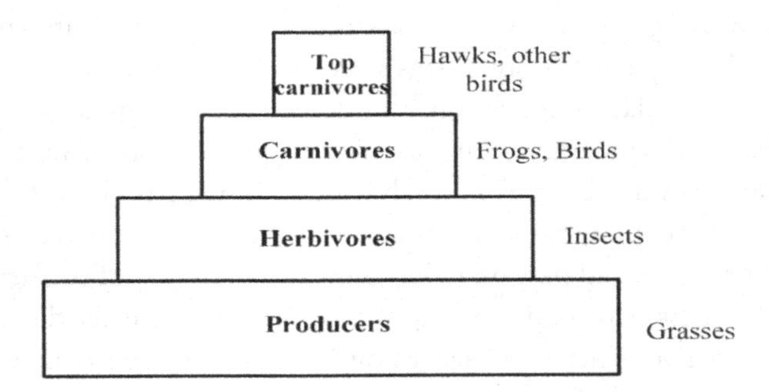

Fig.12: Pyramid of biomass (Grassland ecosystem)

ii. **Pyramid of biomass:** The biomass of the members of the food chain present at any one time forms the pyramid of the biomass (Fig.12). Pyramid of biomass indicates decrease of biomass in each trophical level from base to apex. For example, the total biomass of the producers ingested by herbivores is more than the total biomass of the herbivores in an ecosystem. Likewise, the total biomass of the primary carnivores (or secondary consumer) will be less than the herbivores and so on. The pyramid of biomass can also be upright or inverted. The pyramid of biomass in grassland ecosystem is upright but, the pyramid of biomass in pond ecosystem is inverted. The total biomass of producers (phytoplanktons) is much less as compared to herbivores (zooplanktons, insects), carnivores (small fish) and tertiary carnivores (big fish). Thus, the pyramid takes an inverted shape with narrow base and broad apex (Fig.13).

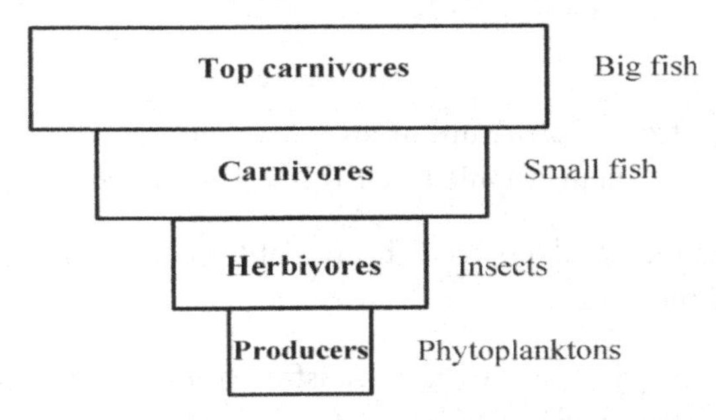

Fig.13: Inverted pyramid of biomass (Pond ecosystem)

iii. **Pyramid of energy:** When production is considered in terms of energy, this pyramid indicates not only the amount of energy flow at each level, but more importantly, the actual role the various organisms played in the transfer of energy (Fig.14). The base upon which the pyramid of energy is constructed is the quantity of organisms produced per unit time, or in other words, the rate at which food material passes through the food chain. Some organisms may have a small biomass, but the total energy they assimilate and pass on, may be considerably greater than that of organisms with a much larger biomass. Energy pyramids are always upright because less energy is transferred from each level than was paid into it. In cases such as in open water communities the producers have less bulk than consumers but the energy they store and pass on must be greater than that of the next level. Otherwise the biomass that producers support could not be greater than that of the producers themselves. This high energy flow is maintained by a rapid turnover of individual plankton, rather than an increase of total mass.

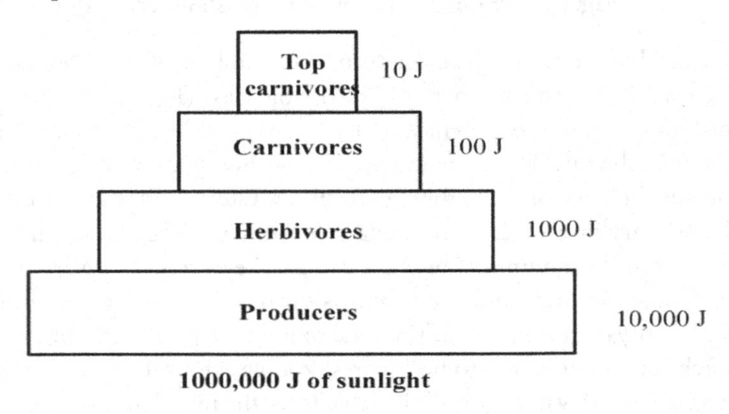

Fig.14: Pyramid of energy

Ecological efficiency:

It is clear from the trophic structure of an ecosystem that the amount of energy decreases at each subsequent trophic level. This is due to two reasons:

1. At each trophic level a part of the available energy is lost in respiration or used up in metabolism.

2. A part of energy is lost at each transformation, *i.e.*, when it moves from lower to higher trophic level as heat.

The ratio of net productivity at one trophic level to the net productivity at previous trophic level is called ecological efficiency. In simple terms, it is the percentage of energy (as biomass) transferred from one trophic level to the next level. Lindman in 1942 defined these ecological efficiencies for the first time and proposed 10% rule *e.g.* if autotrophs produce 100 cal. herbivores might be able to store 10 cal. and carnivores 1 cal. However, there may be slight variations in different ecosystems and ecological efficiencies may range from 5 to 35%.

7.4. Types of ecosystem

Broadly, ecosystems are classified into two types: 1. Aquatic ecosystem and 2. Terrestrial ecosystem. Aquatic ecosystem consists of marine and fresh water and estuary ecosystem. Terrestrial ecosystem consists of forest, grassland, desert and cropland ecosystem. The characteristics of different ecosystems are detailed below:

1. Aquatic ecosystem:

- » In aquatic ecosystems, nutrients are available in soluble form.
- » Here, producers are autotrophs (both algae and aquatic plants) and phytoplanktons which are suspended in water. Consumers are fishes, amphibians and other aquatic animals.
- » They follow the typical inverted pyramid of biomass.
- » Aquatic ecosystem comprises of marine and fresh water ecosystem

i. **Marine ecosystem:** These are gigantic reservoirs of water covering more than 70% of our earth's surface and play a key role in the survival of about 2,50,000 marine species, serving as food for humans and other organisms, give a huge variety of sea products and drugs. Oceans are the major sinks of carbon dioxide and play an important role in regulating many biogeochemical cycles and hydrological cycle, thereby regulating the earth's climate.

ii. **Fresh water ecosystem:** They are again two types:

a. Lentic: ecosystems in standing water like ponds, lakes, swamps and marshy areas

b. Lotic: ecosystems in running water like streams, rivers and canals

The problem with fresh water ecosystem is excessive growth of aquatic weeds increases the BOD thereby disturbing the food chain.

iii. **Estuary ecosystem:**

- » Estuary is a partially enclosed coastal area at the mouth of a river where fresh water and salty seawater meet.

- These are the transition zones which are strongly affected by tidal action.
- Constant mixing of water stirs up the silt which makes the nutrients available for the primary producers.
- The organisms present in estuaries show a wide range of tolerance to temperature and salinity.
- Such organisms are known as eurythermal and euryhaline. Coastal bays and tidal marshes are examples of estuaries.
- Estuary has a rich biodiversity and many of the species are endemic.
- There are many migratory species of fishes like eels and salmons in which half of the life is spent in fresh water and half in salty water. For them estuaries are ideal places for resting during migration, where they also get abundant food.
- Estuaries are highly productive ecosystems.
- The river flow and tidal action provide energy for estuary thereby enhancing its productivity.
- Estuaries are of much use to human beings due to their high food potential.
- However, these ecosystems need to be managed judiciously and protected from pollution.

2. Terrestrial ecosystem:

- Green plants are the main producers in terrestrial ecosystem.
- Biodiversity of terrestrial ecosystem is dynamic and variable both in time and space.
- Air circulation is very rapid and the merely constant percentage of gases is maintained in the ecosystem.
- Soil is highly variable in type, nutrient status and fertility.
- Climate is the major factor in deciding the type of vegetation and ecosystem.
- The vegetation is also decided by the moisture conditions

  - Mesic → vegetation grown in saturated soil
  - Xeric → vegetation grown in limited moisture

» Sometimes, the vegetation will get itself adapted to the vulnerable conditions of climate

  - Halophytes alone can grow near coastal areas

» Macroconsumers are insects, herbivores, carnivores and omnivores. Microconsumers include bacteria, fungi and actinomycetes and they play an active role in biogeochemical cycles.

» Problems of terrestrial ecosystem are: scarcity of water to meet the evapotranspirational needs and climatic vagaries like cyclones, extreme temperature, *etc.*

» Different terrestrial ecosystems are:

i. Forest ecosystem:

» Source of energy in forest ecosystem is solar radiation and biogeochemical cycles.

» In some forest ecosystems (dense and boreal forests), light is the limiting factor.

» Soils are rich in organic matter because of continuous accumulation of litter and soils have good water holding capacity.

» Water is not a limiting factor for soil flora and fauna.

» Microclimate is also congenial for the growth of flora and fauna.

The abiotic environment of forest ecosystem includes the nutrients present in the soil in forest floor which is usually rich in dead and decaying organic matter.

» **Producers**: Producers are mainly big trees, some shrubs and ground vegetation.

» **Primary consumers**: Primary consumers are insects like ants, flies, beetles, spiders, and big animals like elephants, deer, squirrels etc.

» **Secondary consumers**: Secondary consumers are carnivores like snakes, lizards, foxes, birds etc.,

- **Tertiary consumers**: Tertiary consumers are animals like tiger, lion etc.
- **Decomposers:** Decomposers are bacteria and fungi which are found in soil on the forest floor. Rate of decomposition in trophical or sub-trophical forests is more rapid than that in the temperate zones.

ii. Grassland ecosystem:

- The grassland ecosystem occupies about 10% of the earth's surface.
- In grassland ecosystem linear food chains are most common.
- Producers are more in number and there is less competition between the trophic levels.

The abiotic environment includes nutrient like nitrates, sulphates or phosphates and trace elements present in the soil, gases, like CO_2 present in the atmosphere and water etc.

- **Producers:** Producers are mainly grass and some herbs, shrubs, and few scattered trees.
- **Primary consumers**: Primary consumers are grazing animals such as cow, sheep, deer, house, kangaroo, etc. Some insects and spiders have also been included as primary consumers.
- **Secondary consumers**: Secondary consumers are animals like fox, jackals, snakes, lizards, frogs and birds etc.
- **Tertiary consumers**: Tertiary consumers are lion, tiger, eagle, etc.
- **Decomposers** are bacteria, moulds and fungi, like *Penicillium, Aspergillus, etc.* The minerals and other nutrients are thus brought back to the soil and are made available to the producers.

iii. Desert ecosystem:

- Vegetation in desert ecosystem is very sparse and mostly thorny and CAM type.
- The average rainfall is less than 25 cm and water is scarce resource and droughts are very common.

- » Atmosphere is very dry with high wind velocity and low relative humidity.
- » Subjected to temperature extremities *i.e.*, winters are very cold and scorching heat during summer.
- » Soils are light, shallow and highly prone to wind erosion.
- » Soils have very low water holding capacity and the water stored will exhaust within no time.
- » Dust storms are common in desert areas and often lead to extension of desert.
- » Intra specific competition is more at trophic level.
- » In recent days, the vegetation is being modified by the man to overcome the vulnerability of climate.

The abiotic environment of a desert ecosystem includes water which is scarce, soil and climatic factors.

- » **Producers:** The chief producers are shrubs, bushes and some trees whose roots are very extensive and stems and leaves are modified to store water and to reduce loss of water as a result of transpiration. Eg., *Accacia senegal, Prosophis julifera, Zizyphus sp*
- » **Primary consumers**: Primary consumers are insects and small desert animals which get water from succulent plants. They do not drink water even if it is freely available. Camel is also a primary consumer of the desert.
- » **Secondary consumers**: Secondary consumers are carnivores like reptiles having impervious skin which minimize loss water from the surface of body.
- » **Tertiary consumers**: The tertiary consumers are mainly birds which conserve water by excreting solid uric acid.
- » **Decomposers:** Decomposers are bacteria and fungi which can thrive in hot climate conditions. Because of scarcity of flora and fauna, the dead organic matter available is much less and therefore decomposers are also less in number.

iv. Cropland ecosystem:

- » It is not a natural ecosystem, it is an artificial agro-ecosystem

- Involvement of man in this ecosystem is ultimate and hence also known as manmade/man engineered ecosystem.
- Intervention by man is very high to produce more crop with less input for feeding the growing population. Almost all the needs of man are met by this cropland ecosystem.
- In cropland ecosystem, along with ecological factors, socio-economic factors are also considered.
- Physico-chemical environment of the soil is manipulated by addition of organic matter, nutrients to improve the fertility condition of the soil.
- Other interventions of man to the cropland ecosystem are: tillage, fertilizer application, irrigation, weed management, pest and disease management, *etc.*
- Most of the time, the cropland ecosystem is composed of a mono-specific community (mono-cropping).
- Productivity of the crop ecosystem differs depending on the type of farming *i.e.*, rainfed farming, irrigated farming. Productivity of rainfed ecosystem is always less than irrigated ecosystem.
- Tropical agro-ecosystems are usually more productive than temperate agro-ecosystems.
- The problems associated with crop ecosystem are: monocropping, intensive agriculture, fertilizer and pesticide pollution

The abiotic component includes both climatic factors (light, rainfall, temperature, wind, relative humidity) and edaphic factors (soil type, soil pH, water holding capacity, soil fertility, and microbial load).

- **Producers:** The main producers are crops (cereals, pulses, oilseeds, sugar crops, vegetable crops, plantation crops, *etc.*). Weeds associated with crops are also considered as producers.
- **Consumers:** Man, cattle, pests like insects, pathogens, rats, birds, *etc.*
- **Decomposers:** These are microbes present in soil as well as air that decompose the dead organic matter of plants and animals. These are chiefly bacteria, actinomycetes and fungi, responsible for decay, decomposition and humification, making the minerals available again to the producers.

Difference between natural and agro-ecosystem

Natural ecosystem	Agro-ecosystem/Crop ecosystem
» There is more biotic diversity and less synchronization in growth	» Diversity is less, more synchronization in growth and maturity owing to monocultures
» Trophic structure is complex	» Trophic structure is simple
» Less vulnerable to catastrophic changes	» More vulnerable to catastrophic changes
» Less influenced by weather changes	» More influenced by weather
» Pest and disease outbreak are uncommon, all populations being in balance	» Pest and disease outbreaks are of common occurrence
» They are self perpetuating and more stable	» They are controlled by man through artificial practices and are less stable
» They are continuous in space and time and vegetation is naturally selected	» They are discontinuous in space and time and vegetation is selected by man
» They are of long duration, usually years and are of permanent nature	» They are of short duration, many times only of 3-4 months or a year or so and are of temporary nature
» Natural balance exists	» Natural balance does not exist
» Moderately productive	» Highly productive
» Herbivore pressure is more	» Herbivore pressure is less
» They are more mature from succession point of view	» They are not allowed to mature and succession is hated by man

V. Homeostasis:

Ecosystems are capable of maintaining their state of equilibrium. They can regulate their own species structure and functional processes. This capacity of ecosystem of self regulation is known as homeostasis. In ecology the term applies to the tendency for a biological system to resist changes. For example, in a pond ecosystem if the population of zoo-plankton increased, they would consume large number of the phytoplankton and as a result soon zooplankton would be short supply of food for them. As the number zooplankton is reduced because of starvation, phytoplankton population starts increasing. After some time the population size of zooplankton also increases and this process continues at all the trophic levels of the food chain. Note that in a homeostatic system, negative feedback mechanism is responsible for maintaining stability in an ecosystem. However, homeostatic capacity of ecosystems is not unlimited as well as not everything in an ecosystem is always well regulated. Humans are the greatest source of disturbance to ecosystems (Fig.15).

Large number of phytoplankton

↓

Increase population of zooplankton due to excess food available

↓

Reduction in phytoplankton

↓

Population of zooplanktons decrease due to starvation

↓

Population of phytoplankton starts increasing due to less consumption

Fig.15: Homeostasis in ecosystem

Chapter - 8

Bio-Geochemical Cycles

In ecosystems flow of energy is linear but that of nutrients is cyclical. This is because energy flows down hill *i.e.*, it is utilized or lost as heat as it flows forward. The nutrients on the other hand cycle from dead remains of organisms into the soil which are absorbed again from the soil by the roots of green plants and passed on to herbivores and then carnivores. The nutrients locked in the dead remains of organisms and released back into the soil by detrivores and decomposers. This recycling of the nutrients is called biogeochemical or nutrient cycle (*Bio* = living; *geo* = rock; *chemical* = element). There are more than 40 elements required for the various life processes by plants and animals. The entire earth or biosphere is a closed system *i.e.*, nutrients are neither imported nor exported from the biosphere.

There are two important components of a bio-geochemical cycle

1. **Reservoir pool:** Atmosphere or lithosphere, which stores large amounts of nutrients.
2. **Cycling pool or compartments of cycle**: They are relatively short storages of elements/nutrients in the form of plants and animals.

1. Carbon cycle:

The source of all carbon is carbon dioxide present in the atmosphere. It is highly soluble in water; therefore, oceans also contain large quantities of dissolved carbon dioxide (Fig.16). The global carbon cycle consists of following aspects:

Photosynthesis: Green plants in the presence of sunlight utilize CO_2 in the process of photosynthesis and convert the inorganic carbon into organic matter (food) and release oxygen. A part of the food synthesized through photosynthesis is used by plants for their own metabolism and the rest is stored as their biomass which is available to various herbivores, heterotrophs, including human beings and microorganisms as food. Annually 4-9 $x10^{13}$ kg of CO_2 is fixed by green plants of the entire biosphere. Forests acts as reservoirs of CO_2 as carbon fixed by the trees remain stored in them for long period due to their long life cycles. A very large amount of CO_2 is released through forest fires.

Respiration: Respiration is carried out by all living organisms. It is a metabolic process where food is oxidized to liberate energy, CO_2 and water. The energy released from respiration is used for carrying out life processes by living organism (plants, animals, decomposers etc.). Thus, CO_2 is released into of the atmosphere through this process.

Decomposition: All the food assimilated by animals or synthesized by plant is not metabolized by them completely. A major part is retained by them as their own biomass which becomes available to decomposers on their death. The dead organic matter is decomposed by microorganisms and CO_2 is released into the atmosphere by decomposers.

Combustion: Burning of biomass releases carbon dioxide into the atmosphere.

Impact of human activities: The global carbon cycle has been increasingly disturbed by human activities particularly since, the beginning of industrial era. Large scale deforestation and ever growing consumption of fossil fuels by growing numbers of industries, power plants and automobiles are primarily responsible for increasing emission of carbon dioxide. Carbon dioxide has been continuously increasing in the atmosphere due to human activities such as industrialization, urbanization and increasing use and number of automobiles. This is leading to increase concentration of CO_2 in the atmosphere, which is a major cause of global warming.

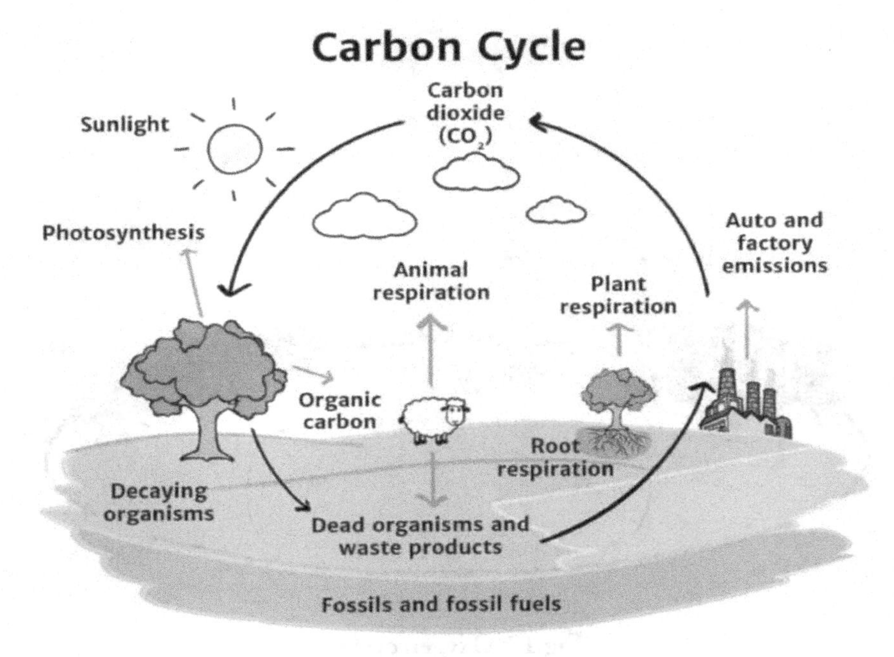

Fig.16: Carbon cycle

2. Oxygen cycle:

Circulation of oxygen in various forms through nature is called oxygen cycle. Oxygen is free in the air and dissolved in water, oxygen is second only to nitrogen in abundance among uncombined elements in the atmosphere. Plants and animals use oxygen to respire and return it to the air and water as carbon dioxide (CO_2). CO_2 is then taken up by algae and terrestrial green plants and converted into carbohydrates during the process of photosynthesis, oxygen being a by-product. The waters of the world are the main oxygen generators of the biosphere; their algae are estimated to replace about 90 percent of all oxygen used. Oxygen is involved to some degree in all the other biogeochemical cycles. For example, over time, detritus from living organisms transfers oxygen-containing compounds such as calcium carbonates into the lithosphere. Despite the burning of fossil fuel and the reduction of natural vegetation (on land and in the sea), the level of atmospheric oxygen appears to be relatively stable because of the increase in plant productivity resulting from agricultural advances worldwide (Fig.17).

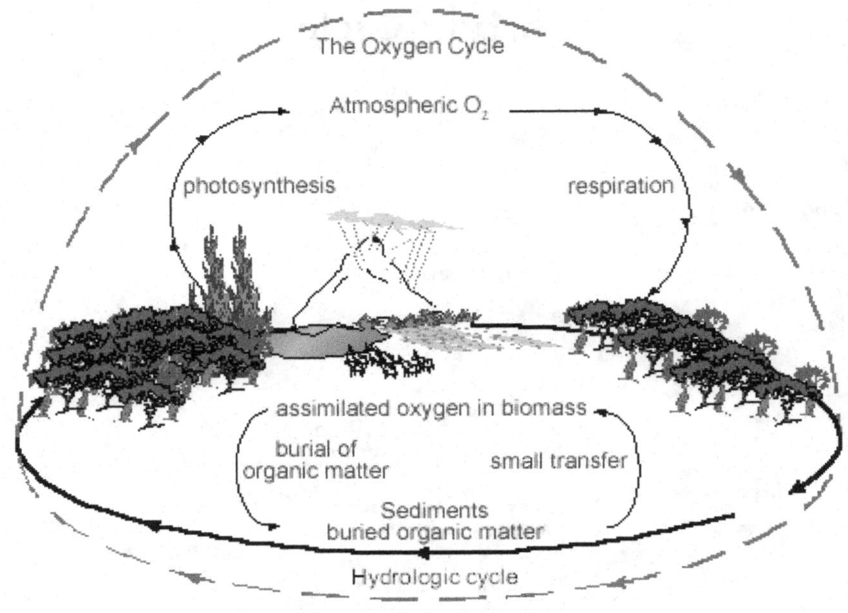

Fig.17: Oxygen cycle

3. Water cycle:

Water is essential for life. No organism can survive without water. Precipitation (rain, snow, slush dew etc.) is the only source of water on the earth. Water received from the atmosphere on the earth returns back to the atmosphere as water vapour resulting from direct evaporation and through evapo-transpiration the continuous movement of water in the biosphere is called water cycle (hydrological cycle). We have already studied that earth is a watery planet of the solar system, about 2/3rd of earth surface is covered with water. However a very small fraction of this is available to animals and plants.

Water is not evenly distributed throughout the surface of the earth. Almost 95% of the total water on the earth is chemically bound to rocks and does not cycle. Out of the remaining 5%, nearly 97.3% is in the oceans and 2.1% exists as polar ice caps. Thus only 0.6% is present as fresh water in the form of atmospheric water vapours, ground and soil water.

The driving forces for water cycle are 1) solar radiation and 2) gravity .

Evaporation and precipitation are two main processes involved in water cycle. These two processes alternate with each other

Water from oceans, lakes, ponds, rivers and streams evaporates by sun's heat energy. Plants also transpire huge amounts of water (Fig.18). Water remains in the vapour state in air and forms clouds which drift with wind. Clouds meet with the cold air in the mountainous regions above the forests and condense to form rain precipitate which comes down due to gravity. On an average 84% of the water is lost from the surface through oceans by evaporation. While 77% is gained by it from precipitation. Water runoff from lands through rivers to oceans makes up 7% which balances the evaporation deficit of the ocean. On land, evaporation is 16% and precipitation is 23%.

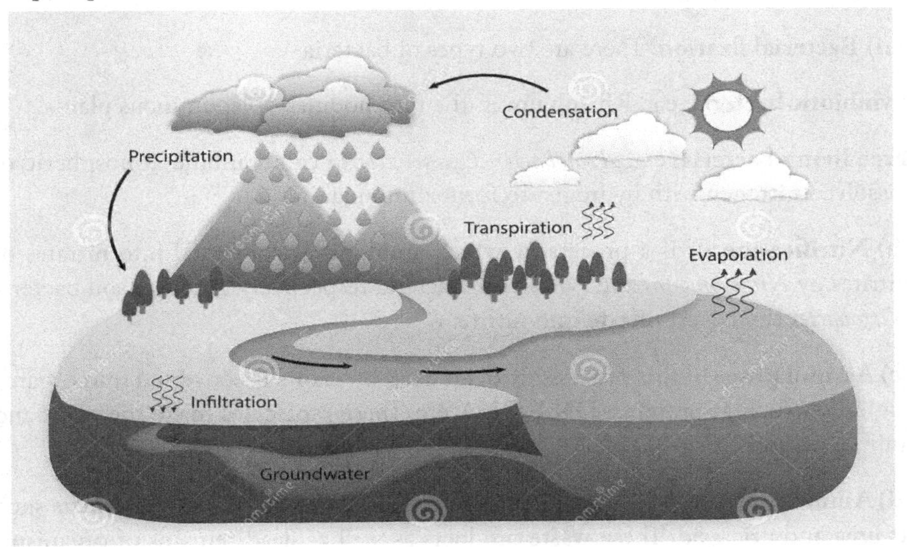

Fig.18: Water cycle

4. Nitrogen cycle:

Nitrogen is an essential component of protein and required by all living organisms including human beings (Fig.19).

Our atmosphere contains nearly 79% of nitrogen but it cannot be used directly by the majority of living organisms. Broadly like carbon dioxide, nitrogen also cycles from gaseous phase to solid phase then back to gaseous phase through the activity of a wide variety of organisms. Cycling of nitrogen is vitally important for all living organisms. There are five main processes which essential for nitrogen cycle are elaborated below.

(a) Nitrogen fixation: This process involves conversion of gaseous nitrogen into Ammonia, a form in which it can be used by plants. Atmospheric nitrogen can be fixed by the following three methods:-

(i) Atmospheric fixation: Lightening, combustion and volcanic activity help in the fixation of nitrogen.

(ii) Industrial fixation: At high temperature (400°C) and high pressure (200 atm.), molecular nitrogen is broken into atomic nitrogen which then combines with hydrogen to form ammonia.

(iii) Bacterial fixation: There are two types of bacteria-

Symbiotic bacteria: e.g. Rhizobium in the root nodules of leguminous plants.

Free living bacteria: e.g. *Azotobacter, Cyanobacteria* can combine atmospheric or dissolved nitrogen with hydrogen to form ammonia.

(b) Nitrification: It is a process by which ammonia is converted into nitrates or nitrites by *Nitrosomonas* and *Nitrococcus* bacteria respectively. Another soil bacteria *Nitrobacter* can covert nitrate into nitrite.

(c) Assimilation: In this process nitrogen fixed by plants is converted into organic molecules such as proteins, DNA, RNA etc. These molecules make the plant and animal tissue.

(d) Ammonification : Living organisms produce nitrogenous waste products such as urea and uric acid. These waste products as well as dead remains of organisms are converted back into inorganic ammonia by the bacteria This process is called ammonification. Ammonifying bacteria help in this process.

(e) De-nitrification: Conversion of nitrates back into gaseous nitrogen is called de-nitrification. Denitrifying bacteria live deep in soil near the water table as they like to live in oxygen free medium. De-nitrification is reverse of nitrogen fixation.

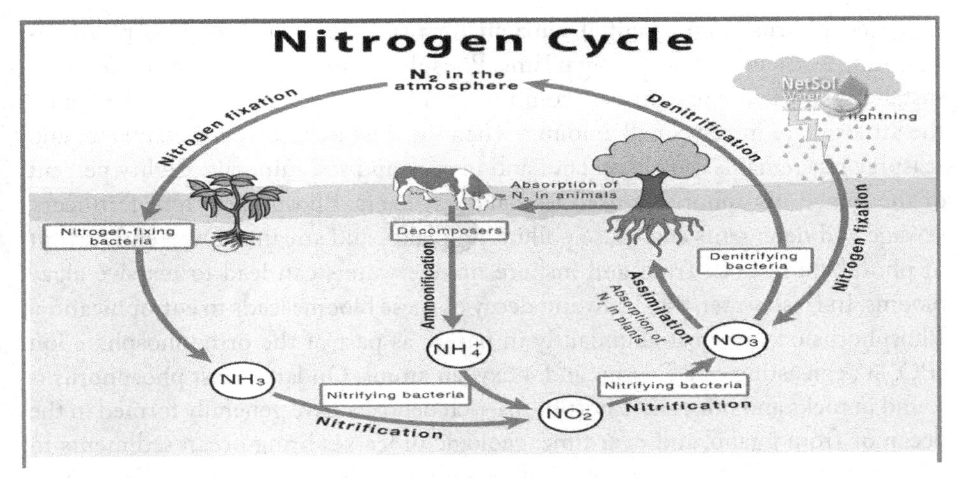

Fig.19: Nitrogen cycle

5. Phosphorous cycle:

The phosphorus cycle describes the movement of phosphorus through the lithosphere, hydrosphere, and biosphere. Unlike many other bio-geochemical cycles, the atmosphere does not play a significant role in the movement of phosphorus, because phosphorus and phosphorus-based compounds are usually solids at the typical ranges of temperature and pressure found on Earth. The production of phosphine gas occurs in only specialized local conditions. Therefore, the phosphorus cycle should be viewed from whole Earth system and then specifically focused on the cycle in terrestrial and aquatic systems (Fig.20).

Living organisms require phosphorus, a vital component of DNA, RNA, ATP, etc, for their proper functioning. Plants assimilate phosphorus as phosphate and incorporate it into organic compounds and in animals, phosphorus is a key component of bones, teeth, etc. On the land, phosphorus gradually becomes less available to plants over thousands of years, since it is slowly lost as runoff. Low concentration of phosphorus in soils reduces plant growth and slows soil microbial growth, as shown in studies of soil microbial biomass. Soil microorganisms act as both sinks and sources of available phosphorus in the biogeochemical cycle. Short-term transformation of phosphorus is chemical, biological, or microbiological. In the long-term global cycle, however, the major transfer is driven by tectonic movement over geologic time.

Humans have caused major changes to the global phosphorus cycle through shipping of phosphorus minerals, and use it for phosphorus fertilizer, and also the shipping of food from farms to cities, where it is lost as effluent.

Phosphorus is an essential nutrient for plants and animals. Phosphorus is a limiting nutrient for aquatic organisms. Phosphorus forms parts of important life-sustaining molecules that are very common in the biosphere. Phosphorus does enter the atmosphere in very small amounts when the dust is dissolved in rainwater and seaspray but remains mostly on land and in rock and soil minerals. Eighty percent of the mined phosphorus is used to make fertilizers. Phosphates from fertilizers, sewage and detergents can cause pollution in lakes and streams. Over-enrichment of phosphate in both fresh and inshore marine waters can lead to massive algae blooms. In fresh water, the death and decay of these blooms leads to eutrophication. Phosphorus occurs most abundantly in nature as part of the orthophosphate ion $(PO_4)3^-$, consisting of a P atom and 4 oxygen atoms. On land most phosphorus is found in rocks and minerals. Phosphorus-rich deposits have generally formed in the ocean or from guano, and over time, geologic processes bring ocean sediments to land. Weathering of rocks and minerals release phosphorus in a soluble form where it is taken up by plants, and it is transformed into organic compounds. The plants may then be consumed by herbivores and the phosphorus is either incorporated into their tissues or excreted. After death, the animal or plant decays, and phosphorus is returned to the soil where a large part of the phosphorus is transformed into insoluble compounds. Runoff may carry a small part of the phosphorus back to the ocean. Generally with time (thousands of years) soils become deficient in phosphorus leading to ecosystem retrogression.

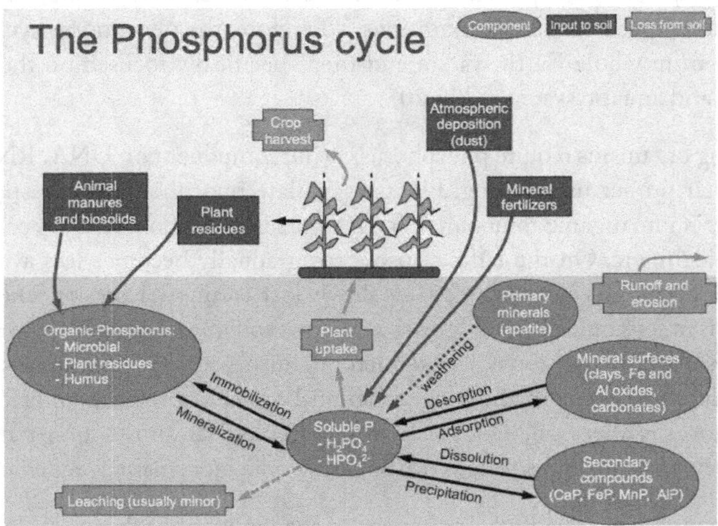

Fig.20: Phosphorus cycle

6. Sulphur cycle:

Circulation of suphur in various forms through nature is called suphur cycle. Suphur occurs in all living matter as a component of certain amino acids. It is abundant in the soil in proteins and, through a series of microbial transformations, ends up as sulfates usable by plants. Suphur-containing proteins are degraded into their constituent amino acids by the action of a variety of soil organisms. The suphur of the amino acids is converted to hydrogen sulfide (H_2S) by another series of soil microbes. In the presence of oxygen, H_2S is converted to suphur and then to suphate by suphur bacteria. Eventually the suphate becomes H_2S. Hydrogen suphide rapidly oxidizes to gases that dissolve in water to form suphurous and suphuric acids. These compounds contribute in large part to the "acid rain" that can kill sensitive aquatic organisms and damage marble monuments and stone buildings (Fig.21).

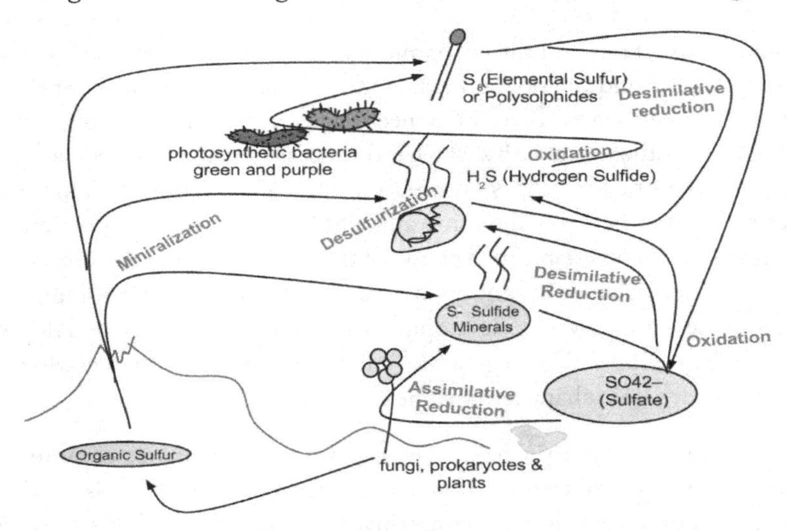

Fig. 21: Sulphur cycle

Chapter - 9

Ecological Succession

Vegetation is hardly stable and thus dynamic, changing over time and space due to variations in climatic and physiographic factors and the activities of the species of the communities themselves. These influences bring about marked changes in the dominants of the existing community, which is thus sooner or later replaced by another community at the same place. This process continues and successive communities develop one after another over the same area until the terminal final community again becomes more or less stable for a period of time. It occurs in a relatively definite sequence. This orderly change in communities is referred as succession. Odum called this orderly process as ecosystem development or ecological succession. Although comparatively less evident than vegetation, animal populations, particularly lower forms also show dynamic character to some extent.

Succession is a natural process by which different groups or communities colonize the same area over a period of time in a definite sequence or it is an orderly process of community development that involves changes in species structure and community processes with time and it is reasonably directional and therefore predictable. Succession is community controlled even though the physical environment determines the pattern.

Causes of Succession

Succession is a series of complex processes, caused by

1. **Initial/initiating cause:** It includes both climatic and biotic elements. Climate includes rainfall, soil erosion, extreme temperature, *etc.* which can destroy the existing vegetation and can lay foundation for succession.

Biotic component includes the organism itself, epidemics (pest and disease outbreaks), *etc.*

2. **Continuous cause:** It includes the situations/conditions which favor the growth and development of new species. *E.g.*, monocropping of sorghum infested once with *Striga* in a small patch will soon leave the entire area infested with *Striga* in a period of time.

3. **Stabilizing cause:** It causes the stabilization of the community. Climate is the chief cause of stabilization and other factors are of secondary value.

Basic types of succession

The various types of succession have been grouped in different ways on the basis of different aspects. Some basic types of succession are however as follows:

i. **Primary succession:** It starts from the primitive substratum where there was no previously any sort of living matter. The first group of organisms establishing there are known as the pioneers, primary community or primary colonizers. The community changes are very slow. *E.g.*, Natural vegetation developed in non-arable, barren, watershed area due to soil and water conservation practices

ii. **Secondary succession:** Another general type of succession is secondary succession which starts from previously built up substrata with already existing living matter. The action of any external force, as a sudden change in climatic factors, biotic intervention, fire, *etc.*, causes the existing community to disappear. Thus, area becomes devoid of living matter but its substratum, instead of primitive, is built up. Such successions are comparatively more rapid.

iii. **Autogenic succession:** After the succession has begun, in most of the cases, it is the community itself which as a result of its reactions with the environment modifies its own environment and thus causing its own replacement by new communities. This course of succession is known as autogenic succession.

iv. **Allogenic succession:** In some cases, however the replacement of the existing community is caused largely by any other external condition and not by the existing organisms. Such a course is referred to as allogenic succession.

On the basis of successive changes in nutritional and energy contents, successions are sometimes classified as:

v. **Autotrophic succession:** It is characterized by early and continued dominance of autotrophic organisms like green plants. It begins in a predominantly inorganic environment and the energy flow is maintained indefinitely. There is gradual increase in the organic matter content supported by energy flow.

vi. **Heterotrophic succession:** It is characterized by early dominance of heterotrophs, such as bacteria, actinomycetes, fungi and animals. It begins in a predominantly organic environment, and there is a progressive decline in the energy content.

In ecological literature, there are still so many other kinds of succession, depending mainly upon the nature of the environment (primarily based upon moisture relations), where the process has begun and thus it may be a hydrosere or hydrarch - starting in regions where water is in plenty such as ponds, lakes, streams, swamp, bog, *etc*; a mesarch - where adequate moisture conditions are present; and a xerosere or xerarch - where moisture is present in minimal amounts, such as dry deserts, rocks, *etc.* Sometimes, there are further distinguished as the lithosere - initiating on rocks; psammosere - on sand and halosere - in saline water or soil.

General process of succession

The whole process of a primary autotrophic succession is actually completed through a number of sequential steps, which follow one another. These steps in sequence are as follows:

1. **Nudation:** This is the development of a bare area without any form of life. The area may develop due to several causes such as landslide, erosion, deposition, or other catastrophic agency. The cause of nudation may be:

   i. **Topographic:** Due to soil erosion by gravity, water or wind, the existing community may disappear. Other causes may be deposition of sand, landslide, volcanic activity and other factors.

   ii. **Climatic:** Glaciers, dry period, hails, storm, frost, and fire *etc.*, may also destroy the community.

   iii. **Biotic:** Man is most important and responsible for destruction of forest grasslands for industry, agriculture, housing etc. Other factors are disease epidemics due to pests which destroy the whole population.

2. **Invasion:** This is the successful establishment of a species in a bare area. The species actually reaches this new site from any other area. This whole process is completed in following three successive stages:

i. **Migration** (dispersal): The seeds, spores or other propagules of the species reach the bare area. This process migration is generally brought about by dispersing agents like air, water, *etc.*

ii. **Ecesis** (establishment): After reaching to new area, the process of successful establishment of the species, as a result of adjustment with the conditions prevailing there, is known as ecesis. In plants, after migration, seeds or propagules germinate, seedlings grow, and adults start to reproduce. Only a few of them are capable of doing this under primitive harsh conditions, and thus most of them disappear. Thus as a result of ecesis, the individuals of species become established in the area.

iii. **Aggregation:** After ecesis, as a result of reproduction, the individuals of the species increase in number and they come close to each other. This process is known as aggregation.

3. **Competition and coaction:** After aggregation of a large number of individuals of the species at the limited place, there develops competition (inter as well as intraspecific) mainly for space and nutrition. Individuals of a species affect each other's life in various ways and this is called coaction. The species, if unable to compete with other species, if present would be discarded. To withstand competition, reproductive capacity, wide ecological amplitude, *etc.* are of much help to the species.

4. **Reaction:** This is the most important stage in succession. The mechanism of the modification of the environment through the influence of living organisms on it is known as reaction. As a result of reactions, changes take place in soil, water, light conditions, temperature of the environment. Due to all these the environment is modified, becoming unsuitable for the existing community which sooner or later is replaced by another community (seral community). The whole sequence of communities that replaces one another in the given area is called a sere and various communities constituting the sere, as seral communities, seral stages or developmental stages. The pioneers are likely to have low nutrient requirements, more dynamic and able to take minerals in comparatively more complex forms. They are small sized and make less demand from environment.

5. **Stabilization (climax):** Finally, there occurs a stage in the process, when the final terminal community becomes more or less stabilized for a longer period of time and it can maintain itself in equilibrium with the climate of the area and it is known as climax. This final community is not replaced, and is known

as climax community and the stage as climax stage. The term climax was coined by Clements (1916).

Theories of Climax

There are three popular theories of climax and they are:

i. **Mono-climax theory (Climatic climax theory):** Mono-climax theory was developed by Clements. This theory recognizes only one climax, determined solely by climate, no matter how great the variety of environment conditions is at the start. According to mono-climax theory, every given region has one climax community towards which all communities are developing. Climax is determined by the regional climate.

ii. **Poly-climax theory:** This theory was proposed by Tansley. This theory considers many different types of vegetation as climatic communities in a given area and they are controlled by number of factors besides climate. The other factors include soil moisture, soil nutrients, topography, slope exposure, fire and animal activity. So, the climax stages may be named depending upon the nature of factor in stabilization. *E.g.*, Topographic climax, biotic climax, edaphic climax, pyric climax, *etc.*

iii. **Climax-pattern hypothesis:** R. H. Whittaker proposed a variation of poly-climax idea, the climax pattern hypothesis. The natural community is adapted to the whole pattern of environmental factors in which it exists. According to this theory, climax communities are patterns of populations varying according to the total environment. Thus there is no discrete number of climax communities and no one factor determines the structure and stability of a climax community.

The monoclimax theory allows for only one climax community in a region, the polyclimax theory allows several climaxes and the climax-pattern hypothesis allows a continuity of climax types varying gradually along environmental gradients and not clearly separable into discrete climax types.

Characteristics of a climax community

» The vegetation is tolerant to environmental conditions

» The vegetation of the climax community will have high ecological amplitude

» They show rich diversity in species composition and a well drained spatial structure

- The community possesses a complex food chain
- Individuals in the climax stage are replaced by the others of the same kind. Thus the species composition maintains equilibrium
- The ecosystem will be balanced and self-sustainable
- There is equilibrium between gross primary production and total respiration
- The energy used from the sunlight and energy released after the decomposition will be balanced
- The uptake of nutrients from the soil and the release of nutrients back to the soil by decomposition will be balanced
- It is an index of the climate of the area. The life or growth forms indicate the climatic type.

Chapter - 10

Competition in Crop Plants

Competition is generally understood or refers to the negative effects on plant growth caused by the presence of neighbors, usually by reducing the availability of resources. Competition can be an important factor controlling plant communities, along with resources, disturbance, herbivory and mutualisms. All plants require a few basic elements, the resources involved are generally light, water, nutrients, space, depending upon the species and the location. The effects of competition are widespread and easily observed in mixtures of crops and managed forests, which is why weeding and thinning are practiced. Competition is also widespread in native habitats, from deserts to wetlands, and is known to have important indeed crucial effects upon recruitment, growth, and reproduction. In the late 1800s, Darwin wrote extensively about the importance of competition in nature, particularly its role in driving natural selection. Thereafter, interest in the phenomenon grew. Many experiments with both crops and wild species were conducted. Models of competitive interactions were also constructed, with the number and size of the models increasing rapidly with the advent of computers in the 1970s.

Competition is ubiquitous among plants and is one of the crucial interactions that influence the health and performance of crops.

Plant competition

Competition can be intra-specific (between members of a species) or inter-specific (between species). In both cases, competition occurs for resources.

Nutrients, water, light and space are the common resources for which plants compete in space and time. While plants obtain light and some essential elements like carbon (C) and oxygen (O) from the air, they get most major and minor nutrients from the soil, such as nitrogen, phosphorus, potassium, calcium, sulphur, magnesium, iron, magnesium, etc. Even the oxygen used for roots is derived from the ground.

Plants gain an advantage if they have better access to nutrients and water by extending their root systems or growing taller and growing more leaves to monopolize a greater amount of light than their neighbors.

There are two main aspects of competition that draws attention of scientists today:

» One chiefly examines the mechanisms of resource competition.

» The second focuses on the reduction in plant fitness that occurs due to limited and shared resources. Moreover, competition can reduce the resources allotted by plants for defense, making them more susceptible to pests and diseases.

Both the mechanisms and results of plant competition are crucial for food production.

In agriculture, the aim is to eliminate competition for resources and to maintain high crop productivity and quality. As monocultures are the norm, the crops are from the same species and variety in a single field. Crop science is an exercise in balancing the need to increase plant density while providing proper spacing, fertilization, irrigation and other agricultural operations to create optimum, competition-free growing conditions for plants.

In open spaces, carbon and oxygen are unlimited in the air. However, in closed atmospheres like greenhouses, even these elementary nutrients have to be considered and supplied.

I. Intra-specific competition

Though tackling the effects of intraspecific competition is central to industrial agriculture, there is surprisingly little work done to quantify the interaction between members of the same cultivar. Most of the studies are focused on interspecific competition of crop and weeds and between intercrops, to a lesser extent. The intraspecific competition effects between plants of the same variety, sown at the same time, can be one or more of the following:

Competition density effect: As the density of plants increases, there is a corresponding decrease in plant performance, reflected in a reduction in the overall size and total productivity of plants.

Size hierarchy development: Competition can affect the relative growth rate of neighboring plants when a few individuals use disproportionate resources. As a result, biomass accumulation among plants can differ. Along with microsite parameters, the number of plants, and the relative time of emergence can determine the position of a plant in the emerging size hierarchy in a field. The size hierarchy effects of competition can assume commercial importance in the case of vegetables that have to meet pre-market size criteria.

Self thinning: Density dependent mortality of plants is unusual in agriculture, as agricultural practices rarely result in densities high enough to lead to mortality. This kind of effect is more common in natural plant communities.

There is also a competition among different organs or sinks of a plant for resource allocation, for example, between vegetative parts and fruits.

II. Inter-specific competition

In agriculture, interspecific competition occurs between weeds and crop plants majorly, and between intercrops of different species, must also be considered. Weeds suppress the growth of young crop plants by shading them and competing for nutrients from the soil and space to grow (Fig. 22). Weeds can reduce the quantity and quality of yield, so it is necessary to minimize crop weed competition by controlling the population of weeds.

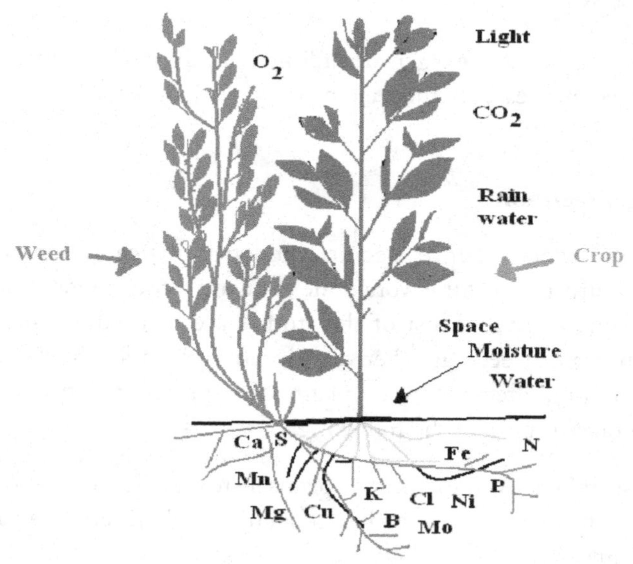

Fig.22: Crop-weed competition

However, crop density and populations can also be optimized to suppress weed growth. The density and biodiversity of weeds will determine their ability to influence crop yield loss.

The competitive ability of various species in an environment depends on many life history factors, such as the following:

- » Time of weed emergence
- » Weed seedling density
- » Morphological and physiological traits, such as leaf area, plant height, root system size, relative biomass, seed size, reproductive strategies, etc.

Parameters to measure plant competition mechanisms

Competition among crop plants is analyzed based on both the mechanisms and outcomes of the relationship. The mechanisms used by crop plants and weeds to get a competitive edge are measured by estimating physiological and morphological parameters.

Leaf area: Leaf area evaluations are essential for measuring competition at three levels: interspecific, intraspecific and within plant resource allocation.

- » Measuring leaf area is typical for crop weed competition, as shading by either species depends on the relative total leaf area. When weeds emerge earlier than crops, they can grow faster and depress crop growth by shading them and limiting access to light.
- » Also, leaf area can indicate plant fitness and photosynthetic potential in intraspecific and interspecific competition, since the final biomass yield depends on leaf area.
- » A balance of resource allocation to vegetative parts measured in leaf area and yield is standard in agriculture.

Several portable instruments like the CI-202 Portable Laser Leaf Area Meter and CI-203 Hand held Laser Leaf Area Meter help take non-destructive, rapid estimations of leaves in the field for single or repeated measurements.

Leaf Area Index: Leaf Area Index (LAI) can also indicate plant health and performance, evaluating light competition in intraspecific and interspecific competition. The amount of light that reaches a plant determines a plant's ability to conduct photosynthesis and, ultimately, crop yield.

A reduction in light interception by plants due to crowding from neighbors and the importance of this reduction can differ based on the crop stage and crop species. LAI is a better indicator than leaf area for light interception as it is the ratio of one-sided leaf area per ground area.

The CI-110 Plant Canopy Imager can measure light interception without an above canopy reference by measuring Photosynthetically Active Radiation (PAR) levels and calculating LAI through the Gap Fraction Method.

Photosynthetic efficiency: Since reducing photosynthesis is one of the mechanisms by which competition affects plant productivity, this physiological process is also directly measured *in- situ.* The photosynthetic rate can evaluate the intraspecific and interspecific competition. Fig. 23 shows that, in two intercropping systems of apple with soybeans and peanuts, photosynthesis decreases in the crops close to apple trees due to light competition.

Photosynthesis is measured in the fields by portable Infrared Gas Analyzers (IRGAs) like the CI-340 Hand held Photosynthesis System, manufactured by CID Bio-Science Inc.

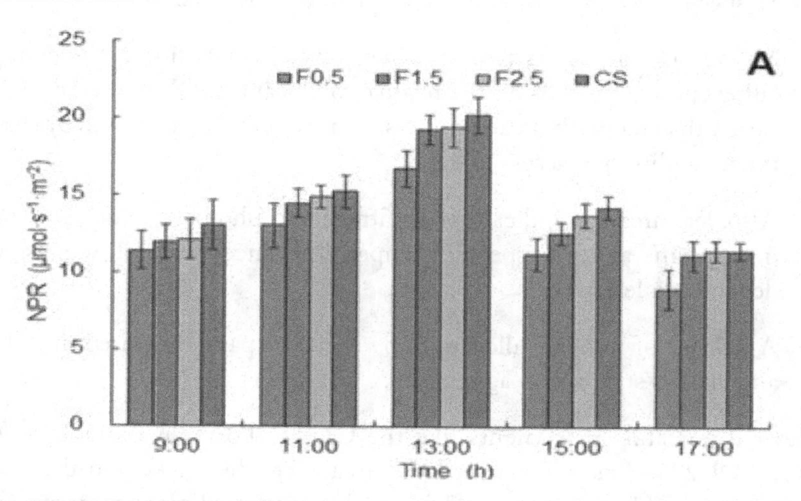

Fig 23: Diurnal variation of net photosynthetic rate (NPR) for the intercropping systems and its control (A. apple + soybean). F0.5, F1.5 and F2.5 were used to represent the sampling points which had different distances (0.5 m, 1.5 m and 2.5 m) from the tree row. Error bars indicate standard deviation."

Root development: Root parameters, such as structure and density, total length, density at different soil depths and effective membrane transporters can provide information on water use efficiency and nutrient acquisition in intraspecific and interspecific competition.

The allocation to root systems can change under competition to improve water and nutrient uptake. The long term underground studies needed to track these changes are now possible with mini-rhizotrons, such as the CI-600 *In-Situ* Root Imager and CI-602 Narrow Gauge Root Imager, which take non-destructive, rapid and accurate scans of root systems through pre installed transparent root tubes.

Nutrient content in plants: The uptake of nitrogen, phosphorus and potassium (NPK), the major limiting nutrients for any plant, can be affected due to competition. The content of these elements can be measured directly in different plant parts through biochemical assays or they can be measured indirectly. For example, nitrogen levels in leaves can be measured via photosynthetic rate.

Yield: The amount of biomass produced by a crop is an intuitive means of understanding how a crop plant or population has been affected by competition. The quantification can be based on plant dry matter or economic yield.

Besides yield, morphological parameters like number of spikes, kernels per spike, test weight, plant height, number of tillers, stem thickness, etc. are also used. These indicators of yield are used in various competition indices to evaluate the results of the competition.

Quality: Competition for resources will impact yield quality, besides the quantity and significantly reduce returns from a farm. These effects are measured by well established quality parameters like soluble solids content, firmness, color and nutritional contents.

Competition Indices

Competition indices are a means to quantify the results of competition. However, these indices do not evaluate the process and mechanisms involved, nor do they track competition progress over time.

Moreover, these competition indices focus on interspecific competition between crops and weeds. There is less information available on intraspecific competition.

Competition indices generally fall into three broad categories:

» Those that seek to **quantify the intensity** of competition, for example, Competitive intensity, and Absolute severity of competition

» Those focused on **determining the effect** of competition, for instance, Relative yield

» Those that **examine the outcome** of competition, for example, Relative reproductive efficiency

Some of the following interspecific competition indices have been revised to evaluate crop intraspecific competition:

» **Relative yield (RY):** Compares crop yield in populations with varying plant densities.

» **Competitive intensity (CI):** Measures plant size in terms of biomass between plants grown at different densities or spacing increments.

» **Absolute severity of competition (ASC):** This index is similar to relative yield, except that one set of plants is grown at no competition and the other is grown at specific densities to compare their yield.

» **Relative reproductive efficiency (RReff):** Plants grown under no/low competition are compared to those at higher densities/competition in seed numbers.

Chapter - 11

Environmental Pollution

'*We spray our elms, and the following spring, trees are silent of robin song, not because we sprayed the robins directly but because the poison traveled step by step through the now familiar elm-earthworm-robin cycle*'.

– Rachael Carson

This quotation appeared in Rachael Carson's book entitled 'Silent Spring'. In the years following the publication of Silent Spring in 1962, the book has inspired controversy and has initiated a major change in thinking about the safety of using pesticides and other toxic chemicals.

Pollution: Undesirable change in the properties of physical, chemical or biological constuents of air, water and soil that affect the life of living organisms or that causes health hazards is called pollution.

This occurs when only short-term economic gains are made at the cost of the long-term ecological benefits by the humanity. No natural phenomenon has led to greater ecological changes than have been made by mankind. During the last few decades we have contaminated mother earth's air, water and land on which life itself depends with a variety of waste products. An average human requires about 12 kg of air each day, which is nearly 12 to15 times greater than the amount of food we eat. Thus even a small concentration of pollutants in the air becomes more significant in comparison to the similar levels present in food. Pollutants that enter water have the ability to spread to distant places especially in the marine ecosystem.

From an ecological perspective pollutants can be classified as follows:

a. **Degradable or non-persistent pollutants:** These can be rapidly broken down by natural processes. Eg: domestic sewage wastes, discarded vegetables, etc.

b. **Slowly degradable or persistent pollutants:** Pollutants that remain in the environment for many years in an unchanged condition and take decades or longer to degrade. Eg: DDT and most plastics.

c. **Non-degradable pollutants:** These cannot be degraded by natural processes. Once they are released into the environment and they are difficult to eradicate and continue to accumulate. Eg: toxic elements like lead or mercury.

1. Air pollution

History of air pollution: The origin of air pollution on the earth can be traced from the times when man started using firewood as a means of cooking and heating. Hippocrates has mentioned air pollution in 400 BC. With the discovery and increasing use of coal, air pollution became more pronounced especially in urban areas. It was recognized as a problem about 700 years ago in London in the form of smoke pollution, which prompted King Edward I to make the first antipollution law to restrict people from using coal for domestic heating in the year 1273. In the year 1300 another act banning the use of coal was passed and defying the law in any form led to imposition of capital punishment. In spite of this air pollution became a serious problem in London during the industrial revolution due to the use of coal in industries. The earliest recorded major disaster was the 'London Smog' that occurred in 1952 that resulted in more than 4000 deaths due to the accumulation of air pollutants over the city for five days.

In Europe, around the middle of the 19th century, a black form of the Peppered moth was noticed in industrial areas. Usually the normal Peppered moth is well camouflaged on a clean lichen covered tree. However the peppered pattern was easily spotted and picked up by birds on the smoke blackened bark of trees in the industrial area, while the black form remained well camouflaged. Thus, while the peppered patterned moths were successful in surviving in clean non-industrial areas, the black coloured moths were successful in industrial areas. With the spread of industrialization, it has been observed that the black forms are not only seen in peppered moth, but also in many other moths. This is a classic case of pollution leading to adaptation.

Air pollution began to increase in the beginning of the 20th century with the development of the transportation systems and large-scale use of fossil fuel based petrol and diesel. The severe air quality problems due to the formation of photochemical smog from the combustion residues of diesel and petrol engines were felt for the first time in Los Angeles. Pollution due to auto-exhaust remains a serious environmental issue in many developed and developing countries including India.

The 'Air Pollution Control Act' in India was passed in 1981 and the 'Motor Vehicle Act' for controlling the air pollution, very recently. These laws are intended to prevent air from being polluted. The greatest industrial disaster leading to serious air pollution took place in Bhopal where extremely poisonous methyl isocyanide gas was accidentally released from the Union Carbide's pesticide manufacturing plant on the night of December 3rd 1984. The effects of this disaster on human health and the soil are felt even today.

Air pollution: Air pollution occurs due to the presence of undesirable solid or gaseous particles in the air in quantities that are harmful to human health and the environment. Air may get polluted by natural causes such as volcanoes, which release ash, dust, sulphur and other gases, or by forest fires that are occasionally naturally caused by lightning.

However, unlike pollutants from human activity, naturally occurring pollutants tend to remain in the atmosphere for a short time and do not lead to permanent atmospheric change. Pollutants that are emitted directly from identifiable sources are produced both by natural events (for example, dust storms and volcanic eruptions) and human activities (emission from vehicles, industries, etc.) are called **primary pollutants**. There are five primary pollutants that together contribute about 90 percent of the global air pollution. These are carbon oxides (CO and CO_2), nitrogen oxides, sulfur oxides, volatile organic compounds (mostly hydrocarbons) and suspended particulate matter.

Pollutants that are produced in the atmosphere when certain chemical reactions take place among the primary pollutants are called **secondary pollutants**. Eg: sulfuric acid, nitric acid, carbonic acid, etc.

Carbon monoxide is a colourless, odorless and toxic gas produced when organic materials such as natural gas, coal or wood are incompletely burnt. Vehicular exhausts are the single largest source of carbon monoxide. The number of vehicles has been increasing over the years all over the world. Vehicles are also poorly maintained and several have inadequate pollution control equipment resulting in release of greater amounts of carbon monoxide. Carbon monoxide is however not a persistent pollutant.

Natural processes can convert carbon monoxide to other compounds that are not harmful. Therefore the air can be cleared of its carbon monoxide if no new carbon monoxide is introduced into the atmosphere.

Sulphur oxides are produced when sulfur containing fossil fuels are burnt. Sulfur dioxide is a colorless pollutant mostly released from industries and power-generating plants. Once this gas in the atmosphere, can be further oxidized to sulfur trioxide (SO_3), which reacts with water vapor or dissolves in water droplets to form sulfuric acid (H_2SO_4) which is the cause of acid rain.

Nitrogen oxides are found in vehicular exhausts. Nitrogen oxides are significant, as they are involved in the production of secondary air pollutants such as ozone.

Bureau of Indian standards assessing the quality of air and fixed the limits for different areas

Air quality standards	Industrial areas	Residential areas	Sensitive areas
	(µg/cubic meter)		
SPM*	500	200	100
SO_2	120	80	30
CO	5000	2000	1000
NO_2	120	80	30

* Suspended Particulate Matter

Hydrocarbons are a group of compounds consisting of carbon and hydrogen atoms. They either evaporate from fuel supplies or are remnants of fuel that did not burn completely. Hydrocarbons are washed out of the air when it rains and run into surface water. They cause an oily film on the surface and do not as such cause a serious issue until they react to form secondary pollutants. Using higher oxygen concentrations in the fuel-air mixture and using valves to prevent the escape of gases, fitting of catalytic converters in automobiles, are some of the modifications that can reduce the release of hydrocarbons into the atmosphere.

Particulates are small pieces of solid material (for example, smoke particles from fires, bits of asbestos, dust particles and ash from industries) dispersed into the atmosphere. The effects of particulates range from soot to the carcinogenic (cancer causing) effects of asbestos, dust particles and ash from industrial plants that are dispersed into the atmosphere. Repeated exposure to particulates can cause them to accumulate in the lungs and interfere with the ability of the lungs to exchange gases.

Types of particulates

Term	Meaning	Examples
Aerosol	General term for particles suspended in air	Sprays from pressurized cans
Mist	Aerosol consisting of liquid droplets	Sulfuric acid mist
Dust	Aerosol consisting of solid particles that are blown into the air or are produced from larger particles by grinding them down	Dust storm
Smoke	Aerosol consisting of solid particles or a mixture of solid and liquid particles produced by chemical reaction such as fires	Cigarette smoke, smoke from burning garbage
Fume	Generally means the same as smoke but often applies specifically to aerosols produced by condensation of hot vapors of metals.	Zinc/lead fumes
Plume	Geometrical shape or form of the smoke coming out of a chimney	
Fog	Aerosol consisting of water droplets	
Smog	Term used to describe a mixture of smoke and fog.	

Lead is a major air pollutant that remains largely unmonitored and is emitted by vehicles. High lead levels have been reported in the ambient air in metropolitan cities. Leaded petrol is the primary source of airborne lead emissions in Indian cities.

Pollutants are also found in indoors from infiltration of polluted outside air and from various chemicals used or produced inside buildings. Both indoor and outdoor air pollution are equally harmful.

A thermal power station consuming 1400 tonnes of coal per day generating 200 MW power also produces:

Pollutant	Emission factor (kg/t of coal)	Emission quantity (t/day)
Aldehydes	0.0025	0.0035
CO	0.25	0.35
Hydrocarbon	0.1	0.14

Oxides of N	10	14
Sulphur oxides	19	13.3
Particulate matter	8	369.6

What happens to pollutants in the atmosphere?

Once pollutants enter the troposphere they are transported downwind, diluted by the large volume of air, transformed through either physical or chemical changes or removed from the atmosphere by rain during which they are attached to water vapour that subsequently forms rain or snow that falls to the earth's surface. The atmosphere normally disperses pollutants by mixing them in the very large volume of air that covers the earth. This dilutes the pollutants to acceptable levels. The rate of dispersion however varies in relation to topography and meteorological conditions.

Effects of air pollution

- Most of the air pollutants are directly involved in the green house effect. The green house gases prevent the heat from being re-radiated to the atmosphere, thereby increasing the atmospheric temperature.
- Exposure to air containing even 0.001 per cent of carbon monoxide for several hours can cause collapse, coma and even death.
- As carbon monoxide remains attached to hemoglobin in blood for a long time, it accumulates and reduces the oxygen carrying capacity of blood. This impairs perception and thinking, slows reflexes and causes headaches, drowsiness, dizziness and nausea.
- Carbon monoxide in heavy traffic causes headaches, drowsiness and blurred vision.
- Sulfur dioxide irritates respiratory tissues. Chronic exposure causes a condition similar to bronchitis.
- Many volatile organic compounds such as (benzene and formaldehyde) and toxic particulates (such as lead, cadmium) can cause mutations, reproductive problems or cancer.
- Deposition of dust and suspended matter on the crop canopy will greatly affect the transpiration and diffusion of gases.

» Suspended particulate matter acts as nuclei for condensation of water vapour to form fog which affects the visibility and penetration of light.

» Oxides cause discoloration of leaves that result in stunted growth and lower productivity of plants.

» At higher concentration of sulphur dioxide majority of the flower buds become stiff and hard. They eventually fall from the plants, as they are unable to flower.

» Pollutants like chlorinated hydrocarbons, CFCs are responsible ozone depletion increasing ultra violet radiation.

Acid rain: When sulphur dioxide and nitrogen oxides are transported by prevailing winds they form secondary pollutants such as nitric acid vapour, droplets of sulfuric acid and particles of sulphate and nitrate salts. These chemicals descend on the earth's surface in two forms: wet (as acidic rain, snow, fog and cloud vapour) and dry (as acidic particles). The resulting mixture is called acid deposition, commonly called acid rain.

Acid deposition has many harmful effects especially when the pH falls below 5.1 for terrestrial systems and below 5.5 for aquatic systems. It contributes to human respiratory diseases such as bronchitis and asthma, which can cause premature death. It also damages statues, buildings, metals and car finishes. Acid deposition can damage tree foliage directly but the most serious effect is weakening of trees so they become more susceptible to other types of damage. The nitric acid and the nitrate salts in acid deposition can lead to excessive soil nitrogen levels. This can over stimulate growth of other plants and intensify depletion of other important soil nutrients such as calcium and magnesium, which in turn can reduce tree growth and vigour.

Management of air pollution: Air pollution can be managed through following measures

» Use of low sulphur coal in industries or removing the sulphur from coal.

» Use of coal should be reduced and alternate sources should be evolved. Stress should be given on exploitation of renewable energy sources like solar and wind energy.

» Legal clearence for industries after proper environmental impact assessment studies.

» Afforestation programmes and planting of green belts around cities and industrial areas. The plants selected should be resistant to dust, sequester more carbon dioxide and should be tolerant to pollution.

» Ex: *Dalbergia sisso, Albezia lebek, Ficus sps, Legostromia speciosa, Casuarina equisetifolia, Tectona grandis, Terminalia arjuna,*

» Reducing green house gas emissions through use of renewable energy sources.

» Adopting the principles of reduce, reuse and recycle.

» Use of energy efficient products like biofuel blends, CNG etc.

» Use of public transport systems.

» Avoid burning of biomass and plastics.

» Vehicular pollution can be checked by regular tune up of engines, replacement of old, more polluting vehicles, installing catalytic converters, by engine modification to become fuel efficient and to reduce CO and NO_2 emission.

» Replacement of fossil fuels with alternate sources of energy.

» Use of lead free fossil fuels.

» One of the effective means of controlling air pollution is to have proper equipment in place. This includes devices for removal of pollutants from the flue gases though scrubbers, closed collection recovery systems through which it is possible to collect the pollutants before they escape, use of dry and wet collectors, filters, electrostatic precipitators, etc.

» Providing a greater height to the stacks can help in facilitating the discharge of pollutants as far away from the ground as possible.

» Industries should be located in places so as to minimize the effects of pollution after considering the topography and the wind directions.

» Substitution of raw material that causes more pollution with those that cause less pollution can be done.

Legal aspects of air pollution control in India

The Air Act (Prevention and Control of Pollution) was legislated in 1981. The Act provided for prevention, control and abatement of air pollution. In areas notified under this Act no industrial pollution causing activity could come up without the permission of the concerned State Pollution Control Board. But this Act was not strong enough to play a precautionary or a corrective role. After the Bhopal disaster, a more comprehensive Environment Protection Act (EPA) was passed in 1986. This Act for the first time conferred enforcement agencies with necessary punitive powers to restrict any activity that can harm the environment. To regulate vehicular pollution the Central Motor Vehicles Act of 1939 was amended in 1989. Following this amendment the exhaust emission rules for vehicle owners were notified in 1990 and the mass emission standards for vehicle manufacturers were enforced in 1991 for the first time. The mass emission norms have been further revised for 2000.

2. Water pollution

Water is the essential element that makes life on earth possible. Without water there would be no life. We usually take water for granted. When the quality or composition of water changes directly or indirectly as a result of man's activities such that it becomes unfit for any purpose it is said to be polluted.

Point sources of pollution: When a source of pollution can be readily identified because it has a definite source and place where it enters the water it is said to come from a **point source.** Eg. Municipal and Industrial discharge pipes. When a source of pollution cannot be readily identified, such as agricultural runoff, acid rain, etc, they are said to be **non-point sources** of pollution.

Causes of water pollution: There are several classes of common water pollutants.

There are **disease-causing agents** (pathogens) which include bacteria, viruses, protozoa and parasitic worms that enter water from domestic sewage and untreated human and animal wastes. Human wastes contain concentrated populations of coliform bacteria such as *Escherichia coli* and *Streptococcus faecalis.* These bacteria are not harmful in low numbers. Large amounts of human waste in water, increases the number of these bacteria which cause gastrointestinal diseases. Other potentially harmful bacteria from human wastes may also be present in smaller numbers. Thus, the greater the amount of wastes in the water the greater is the chances of contracting diseases from them.

Another category of water pollutants is **oxygen depleting wastes.** These are organic wastes that can be decomposed by aerobic (oxygen requiring) bacteria. Large populations of bacteria use up the oxygen present in water to degrade these wastes. In the process this degrades water quality. The amount of oxygen required to break down a certain amount of organic matter is called the **biological oxygen demand** (BOD). The amount of BOD in the water is an indicator of the level of pollution. If too much organic matter is added to the water all the available oxygen is used up. This causes fish and other forms of oxygen dependent aquatic life to die. Thus anaerobic bacteria (those that do not require oxygen) begin to break down the wastes. Their anaerobic respiration produces chemicals that have a foul odour and an unpleasant taste that is harmful to human health.

Another class of pollutants is **inorganic plant nutrients**. These are water soluble nitrates and phosphates that cause excessive growth of algae and other aquatic plants. The excessive growth of algae and aquatic plants due to added nutrients is called eutrophication. They may interfere with the use of the water by clogging water intake pipes, changing the taste and odour of water and cause a buildup of organic matter. As the organic matter decays, oxygen levels decrease and fish and other aquatic species die. The quantity of fertilizers applied in a field is often many times more than is actually required by the plants. The chemicals in fertilizers and plant protection chemicals we use pollute soil and water. While excess fertilizers cause eutrophication, pesticides cause bioaccumulation and biomagnification. Pesticides which enter water bodies are introduced into the aquatic food chain. They are then absorbed by the phytoplanktons and aquatic plants. These plants are eaten by the herbivorous fish which are in turn eaten by the carnivorous fish which are in turn eaten by the water birds. At each link in the food chain these chemicals which do not pass out of the body are accumulated and increasingly concentrated resulting in biomagnification of these harmful substances.

One of the effects of accumulation of high levels of pesticides such as DDT, BHC is that birds lay eggs with shells that are much thinner than normal. This results in the premature breaking of these eggs, killing the chicks inside. Birds of prey such as hawks, eagles and other fish eating birds are affected by such pollution. Although DDT has been banned in India for agricultural use and is to be used only for malaria eradication, it is still used in the fields as it is cheap.

An another class of water pollutants is **water soluble inorganic chemicals** which are acids, salts and compounds of toxic metals such as mercury and lead. High levels of these chemicals can make the water unfit to drink, harm fish and other aquatic life, reduce crop yields and accelerate corrosion of equipment that use this water.

Minamata - An important lesson about Mercury: A case of human mercury poisoning which occurred about 40 years ago in the Minamata bay in Japan taught the world an important lesson about the dangers of mercury poisoning. A large plastics plant located near the Minamata bay used a mercury containing compound in a reaction to produce vinyl chloride, a common plastic material. The left over mercury was dumped into the Bay along with other wastes from the plant. Though the mercury was in its less toxic inorganic state when dumped microorganisms at the bottom of the bay converted the mercury into its organic form. This organic mercury then entered into the tissues of fish which were in turn consumed by the people living in the area. The contaminated fish thus caused an outbreak of poisoning, killing and affecting several people. Mothers who had eaten the contaminated fish gave birth to infants who showed signs of mercury poisoning. Mercury poisoning is thus called Minamata Disease.

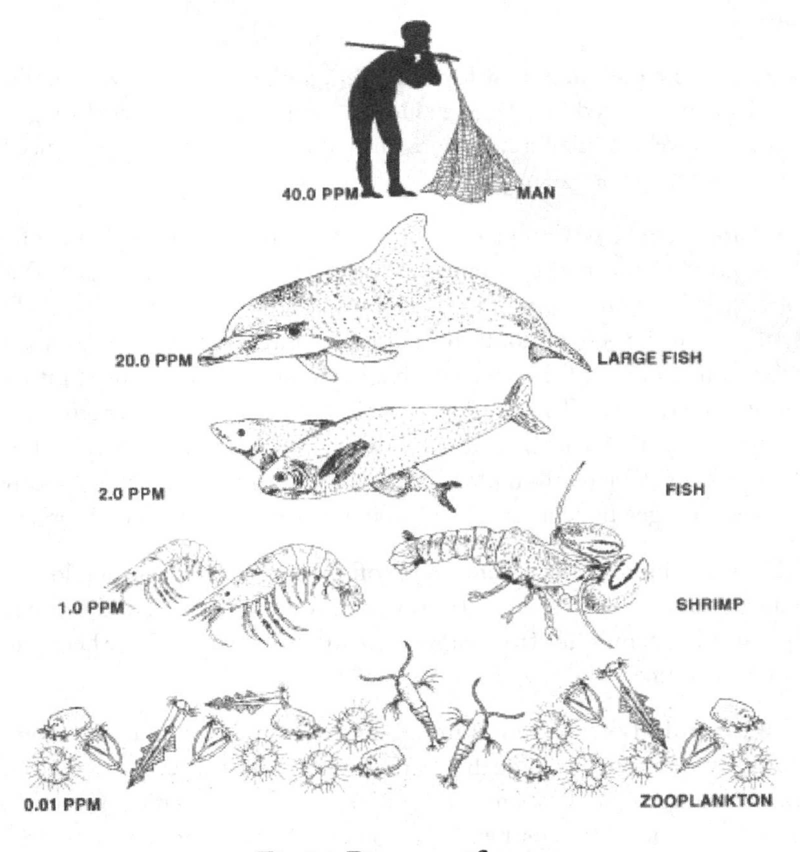

Fig.24: Bio-magnification

Another cause of water pollution is a variety of **organic chemicals**, which include oil, gasoline, plastics, pesticides, cleaning solvents, detergent and many other chemicals. These are harmful to aquatic life and human health. They get into the water directly from industrial activity either from improper handling of the chemicals in industries and more often from improper and illegal disposal of chemical wastes.

Sediments of suspended matter is another class of water pollutants. These are insoluble particles of soil and other solids that become suspended in water. This occurs when soil is eroded from the land. High levels of soil particles suspended in water, interferes with the penetration of sunlight. This reduces the photosynthetic activity of aquatic plants and algae disrupting the ecological balance of the aquatic bodies. When the velocity of water in streams and rivers decreases, the suspended particles settle down at the bottom as sediments. Excessive sediments that settle down destroys feeding and spawning grounds of fish, clogs and fills lakes, artificial reservoirs etc.

Water soluble radioactive isotopes are yet another source of water pollution. These can be concentrated in various tissues and organs as they pass through food chains and food webs. Ionizing radiation emitted by such isotopes can cause birth defects, cancer and genetic damage.

Hot water let out by power plants and industries that use large volumes of water to cool the plant result in rise in temperature of the local water bodies. Thermal pollution occurs when industry returns the heated water to a water source. Power plants heat water to convert it into steam, to drive the turbines that generate electricity. For efficient functioning of the steam turbines, the steam is condensed into water after it leaves the turbines. This condensation is done by taking water from a water body to absorb the heat. This heated water, which is at least 15°C higher than the normal is discharged back into the water body. The warm water not only decreases the solubility of oxygen but changes the breeding cycles of various aquatic organisms.

Oil is washed into surface water in runoff from roads and parking lots which also pollutes groundwater. Leakage from underground tanks is another source of pollution. Accidental oil spills from large transport tankers at sea have been causing significant environmental damage.

Ground water pollution: Surface water pollution is highly visible and often gets a lot of media attention, but a much greater threat to human life comes from our groundwater being polluted which is used for drinking and irrigation. Groundwater flows are slow and not turbulent hence the contaminants are not effectively diluted and dispersed as compared to surface water. Moreover pumping groundwater and treating it is very slow and costly. Hence, it is extremely essential to prevent the pollution of groundwater in the first place.

Ground water is polluted due to:

- » Urban run-off of untreated or poorly treated waste water and garbage into open drains
- » Industrial waste storage located above or near aquifers
- » Agricultural practices such as the application of large amounts of fertilizers and pesticides, etc. in the rural sector
- » Leakage from underground storage tanks containing gasoline and other hazardous substances
- » Leachate from landfills
- » Poorly designed and inadequately maintained septic tanks
- » Mining wastes

Due to excessive use of nitrogenous fertilizers, the concentration of NO_3 (nitrate) in the ground water is increasing due to leaching and even entering into the food chain through drinking water. The microbes present in intestine converts nitrate to nitrite (NO_2). Nitrate combined with hemoglobin of blood forms an abnormal compound called mathehemoglobin. This mathehemoglobin causes respiratory problems and also changes the colour of the skin to blue and hence, it is also known as blue baby syndrome. In a healthy person, mathehemoglobin content is 0.8% and if it reaches 10 % or above then it is a problem. This entire process is also referred as 'nitrate poisoning'. The permissible limit of nitrate in water is <10 mg/lit of water.

Severe cases of arsenic poisoning from contaminated groundwater have been reported from West Bengal in what is known today as the worst case of groundwater pollution. Groundwater arsenic contamination in West Bengal was first reported in December 1983 when 63 people from three villages located in different districts were identified by health officials as suffering from arsenic poisoning.

The state of India's rivers

India has always had a tradition of worshipping rivers. Most of the rivers in India are named after gods, goddesses or saints. However, a large majority of the Indian population including those who worship the rivers do not think twice before polluting a river. Urbanization, industrialization, spiritual practices, excess withdrawal of water, agricultural run-off, improper agricultural practices and various religious and social practices all contribute to river pollution in India. Every single river in India be it the Ganga, Yamuna, Godavari, Cauvery or the Krishna have their own share of problems due to pollution. Waters from the Ganga and the Yamuna are drawn for irrigation through the network of canals as soon as these rivers reach the plains reducing the amount of water that flows downstream. What flows in the river is water from small nalas, and streams that carry with them sewage and industrial effluents. The residual freshwater, is unable to dilute the pollutants and the rivers turn into stinking sewers. In spite of data from scientifically competent studies conducted by the Central Pollution Control Board (CPCB), the Government has not been able to tackle this issue. Sewage and municipal effluents account for 75% of the pollution load in rivers while the remaining 25% is from industrial effluents and non-point pollution sources.

In 1985, India launched the Ganga Action plan (GAP) the largest ever river clean-up operation in the country. The plan has been criticized for, overspending and slow progress. The GAP Phase II executed in 1991 included cleaning operations for the tributaries of the Ganga, ie; the Yamuna, Gomti and Damodar. Thus, the Yamuna Action Plan (YAP), Gomti Action Plan and the Damodar Action plan were added.

In 1995 the National River Conservation plan was launched. Under this all the rivers in India were taken up for clean-up operations. In most of these plans, attempts have been made to tap drains, divert sewage to sewage treatment plants before letting out the sewage into the rivers. The biggest drawback of these river cleaning programs was that they failed to pin responsibilities as to who would pay for running the treatment facilities in the long run. With the power supply being erratic and these plants being heavily dependent on power, most of these facilities lie underutilized. Moreover the problem of river pollution due to agricultural runoff has not been addressed in these programs. NRCP is scheduled to be completed by March 2005. The approved cost for the plan is Rs. 772.08 crores covering 18 rivers in 10 states including 46 towns. The cost is borne entirely by the Central Government and the Ministry of Environment and Forests is the nodal agency that co-ordinates and monitors the plan. Under this plan the major activities include treating the pollution load from sewer systems of towns and cities, setting up of sewage treatment plants, electric crematoria, low cost sanitation facilities, riverfront development, afforestation and solid waste management.

Management of water pollution

» While the foremost necessity is prevention, setting up effluent treatment plants and treating waste through these can reduce the pollution load in the recipient water. The treated effluent can be reused for either gardening or cooling purposes wherever possible.

» A few years ago a new technology called **Root Zone Process** has been developed by Thermax. This system involves running contaminated water through the root zones of specially designed reed beds. The reeds, which are essentially wetland plants have the capacity to absorb oxygen from the surrounding air through their stomatal openings. The oxygen is pushed through the porous stem of the reeds into the hollow roots where it enters the root zone and creates conditions suitable for the growth of numerous bacteria and fungi. These micro-organisms oxidize impurities in the waste waters, so that the water which finally comes out with clean.

» The agricultural sector uses upwards of 70 per cent of the surface water supplies around the earth for everything from livestock production to farming. Unfortunately, **agriculture is one of the primary causes of water pollution**. Whenever it rains, the pesticides and fertilizers wash away with the storm water. To foster the use of green agriculture, consider planting trees and other plants nearby bodies of water, which will keep chemicals from being washed away when it rains. You should also avoid using pesticides that contain harmful chemicals.

One way of reducing the pollution load in waters is through the introduction of sewage treatment plants. This will reduce the biological oxygen demand (BOD) of the final product. The various processes involved in sewage treatment are discussed below.

Primary treatment: Under this treatment plants use physical processes such as screening and sedimentation to remove pollutants that will settle, float or, those are too large to pass through simple screening devices. This includes stones, sticks, rags, and all such material that can clog pipes.

A screen consists of parallel bars spaced 2 to 7 cm apart followed by a wire mesh with smaller openings. One way of avoiding the problem of disposal of materials collected on the screens is to use a device called a comminuter which grinds the coarse material into small pieces that can then be left in the waste water. After screening the wastewater passes into a grit chamber. The detention time is chosen to be long enough to allow lighter, organic material to settle. From the grit chamber the sewage passes into a primary settling tank (also called as sedimentation tank) where the flow speed is reduced sufficiently to allow most of the suspended solids to settle out by gravity. If the waste is to undergo only primary treatment it is then chlorinated to destroy bacteria and control odours after which the effluent is released. Primary treatment normally removes about 35 percent of the BOD and 60 percent of the suspended solids.

Secondary treatment: The main objective of secondary treatment is to remove most of the BOD. There are three commonly used approaches: trickling filters, activated sludge process and oxidation ponds. Secondary treatment can remove at least 85 per cent of the BOD. A trickling filter consists of a rotating distribution arm that sprays liquid wastewater over a circular bed of 'fist size' rocks or other coarse materials. The spaces between the rocks allow air to circulate easily so that aerobic conditions can be maintained. The individual rocks in the bed are covered with a layer of slime, which consists of bacteria, fungi, algae, etc. which degrade the waste trickling through the bed. This slime periodically slides off individual rocks and is collected at the bottom of the filter along with the treated wastewater and is then passed on to the secondary settling tank where it is removed. In the activated sludge process the sewage is pumped into a large tank and mixed for several hours with bacteria rich sludge and air bubbles to facilitate degradation by micro-organisms.

The water then goes into a sedimentation tank where most of the microorganisms settle out as sludge. This sludge is then broken down in an anaerobic digester where methane-forming bacteria slowly convert the organic matter into carbon dioxide, methane and other stable end products. The gas produced in the digester is 60 per cent methane, which is a valuable fuel and can be put to many uses within the treatment plant itself. The digested sludge, which is still liquid, is normally pumped out onto sludge drying beds where evaporation and seepage remove the water. This dried sludge is potentially a good source of manure. Activated sludge tanks use less land area than trickling filters with equivalent performance. They are also less expensive to construct than trickling filters and have fewer problems with flies and odour and can also achieve higher rates of BOD removal. Thus although the operating costs are a little higher due to the expenses incurred on energy for running pumps and blowers they are preferred over trickling filters.

Tertiary treatment: Tertiary treatment of wastewater means final filtration of the treated effluent. When needed, it sometimes involves using alum to remove phosphorus particles from the water. Alum also causes any solids that were not removed by primary and secondary wastewater treatment to group so they can be removed by filters. When necessary, the filters are backwashed to remove the buildup of floc, which allows the filters to continue operating effectively.

Disinfection: An important part of wastewater treatment involves the addition of chlorine to the final effluent before discharge. This process injects chlorine into the headworks of a serpentine effluent detention chamber. Chlorination in wastewater treatment kills bacteria and viruses, and eliminates parasites such as Giardia and Cryptosporidium, which can cause very serious illnesses. In summary, this process disinfects water so that it is safe to reuse or recycle.

Di-chlorination: The final stage of the tertiary wastewater treatment process involves removing the chlorine that was used to disinfect the water. This step is very important because chlorine is harmful to aquatic life. Chlorine also reduces biological water quality when it is present in high concentrations. To remove the chlorine, a compound called sodium bisulfite is added to the water. Chlorine ions in the water react with this chemical and are removed. Once the chlorine concentration has been reduced to a safe level, the treated water is now considered clean enough to be safely released into the environment.

3. Soil pollution

Soil is the basis of agriculture. All crops for human food and animal feed depend on it. Soil is the ultimate recipient of enormous quantity of man made waste products. Industrial wastes, sewage, sludge and polluted water cause soil pollution.

Sources of soil pollution:

The sources which pollute the soil are of two fold:

a) Agricultural sources: Animal wastes, pesticides, fertilizers etc.

b) Non agricultural wastes: sewage water, industrial effluents etc.

Major soil pollutants: Materials that find their entry into the soil system having long persistence and accumulate in toxic concentrations are called soil pollutants. There are four types of soil pollutants.

a. **Inorganic toxic compounds:** Inorganic residues in industrial waste pose serious problem of disposal. They contain heavy metals which have potential toxicity. Among heavy metals nickel, cadmium and lead are having high toxic potential, zinc and mercury have moderate and chromium has low potential. Arsenic compounds which are used in insecticidal preparations are also having high toxic potential.

b. **Organic wastes:** Organic wastes of various types cause pollution hazards. Domestic garbage, municipal sewage and industrial wastes when left in heaps or improperly disposed seriously affect the health of human beings, plants and animals. Sewage contains detergents, water softeners, borates, phosphates and other salts in large amounts.

c. **Organic pesticides:** Pesticide persistence in soil and movement into water streams may also lead to their entry into food and create health problems (Bio-magnification). Pesticides particularly aromatic organic compounds are not degraded rapidly and therefore have long persistence time. The persistence time of some of chemicals are as below.

Pesticide	**Persistence time**
Chlordane	12 years
BHC	11 years
DDT	10 years
Aldrin	9 years
2,3,6 TBA	2-5 years
Diuron	16 months
Simazine	17 months
2,4-D	2-8 weeks

The main method of checking pesticides pollution is to increase the organic matter content of the soil and choose such pesticides which are non persistent and leave no harmful residues in soil and environment.

d. **Radioactive wastes:** Wastes from atomic reactors and nuclear power plants containing different kinds of isotopes which are most dangerous. These wastes affect aquatic plants and animals also. They cause ionization of various body fluids, gene mutations and chromosomal mutations.

Management of soil pollution:

» Sanitary landfills are to be promoted in place of open dumping

» Proper solid waste management, reduce, recycling and reuse of materials like paper, metals, hospitals, slaughter houses and glass is to be encouraged

» Organic wastes if treated properly, becomes a good quality compost

» Proper management of agricultural land and the practice of organic farming

» Use of pesticides and fertilizers should be minimized

» **Bio-remediation:** Bioremediation is a branch of biotechnology that employs the use of living organisms, like microbes and bacteria, in the removal of contaminants, pollutants, and toxins from soil, water and other environments.

The emerging science and technology of bioremediation offers an alternative method to detoxify contaminants. Bioremediation has been demonstrated and is being used as an effective means of mitigating hydrocarbons, halogenated organic solvents, halogenated organic compounds, non-chlorinated pesticides and herbicides, nitrogen compounds, metals (lead, mercury, chromium), radio-nuclides.

Bioremediation which occurs without human intervention and other than monitoring is often called natural attenuation. This natural attenuation relies on natural conditions and behavior of soil microorganisms that are indigenous to soil. Biostimulation also utilizes indigenous microbial populations to remediate contaminated soils. Biostimulation consists of adding nutrients and other substances to soil to catalyze natural attenuation processes. Bioaugmentation involves introduction of exogenic microorganisms (sourced from outside the soil environment) capable of detoxifying a particular contaminant, sometimes employing genetically altered microorganisms (Biobasics, 2006).

During bioremediation, microbes utilize chemical contaminants in the soil as an energy source and through oxidation-reduction reactions, metabolize the target contaminant into useable energy for microbes. By-products (metabolites) released back into the environment are typically in a less toxic form than the parent contaminants. For example, petroleum hydrocarbons can be degraded by microorganisms in the presence of oxygen through aerobic respiration. The hydrocarbon loses electrons and is oxidized while oxygen gains electrons and is reduced. The result is formation of carbon dioxide and water. When oxygen is limited in supply or absent, as in saturated or anaerobic soils or lake sediment, anaerobic (without oxygen) respiration prevails. Generally, inorganic compounds such as nitrate, sulfate, ferric iron, manganese, or carbon dioxide serve as terminal electron acceptors to facilitate biodegradation.

Three primary ingredients for bioremediation are: 1) presence of a contaminant, 2) an electron acceptor, and 3) presence of microorganisms that are capable of degrading the specific contaminant. Generally, a contaminant is more easily and quickly degraded if it is a naturally occurring compound in the environment, or chemically similar to a naturally occurring compound, because microorganisms capable of its biodegradation are more likely to have evolved. Petroleum hydrocarbons are naturally occurring chemicals; therefore, microorganisms which are capable of attenuating or degrading hydrocarbons exist in the environment.

In situ bioremediation causes minimal disturbance to the environment at the contamination site. In addition, it incurs less cost than conventional soil remediation or removal and replacement treatments because there is no transport of contaminated materials for off-site treatment. However, *in situ* bioremediation has some limitations: 1) it is not suitable for all soils, 2) complete degradation is difficult to achieve, and 3) natural conditions (i.e. temperature) are hard to control for optimal biodegradation.

Ex situ bioremediation, in which contaminated soil is excavated and treated elsewhere, is an alternative. *Ex situ* bioremediation approaches include use of bioreactors, land farming, and biopiles. In the use of a bioreactor, contaminated soil is mixed with water and nutrients and the mixture is agitated by a mechanical bioreactor to stimulate action of microorganisms. This method is better-suited to clay soils than other methods and is generally a quick process. Land farming involves spreading contaminated soil over a collection system and stimulating microbial activity by allowing good aeration and by monitoring nutrient availability. Biopiles are mounds of contaminated soils that are kept aerated by pumping air into piles of soil through an injection system.

» **Chemical method:** Chemical methods aim at addition of chemicals or solvents into the polluted soils so as to stabilize the pollutants and convert them into less toxic forms that are harmless to the water bodies, plants and human beings. Since, complete soil remediation is difficult to achieve with biological methods alone hence, the amalgamation of both biological and chemical methods has gained much attention of the scientists. Besides that, the harmful effects of the use of chemical methods should also be considered before implementing on a pilot-scale. The materials generally used for chemical treatments are metallic oxides, clays or biomaterials. Remediation efficiency of these materials depends upon the soil texture, organic matter present in the soil, type of metal contaminant etc. Furthermore, chemical methods can offer a fast remediation compared to slow bioremediation process.

4. Noise pollution

Noise may not seem as harmful as the contamination of air or water but it is a pollution problem that affects human health and can contribute to a general deterioration of environmental quality.

Noise is undesirable and unwanted sound. Not all sound is noise. What may be considered as music to one person may be noise to another. It is not a substance that can accumulate in the environment like most other pollutants. Sound is measured in a unit called the 'Decibel'.

Decibel levels of common sounds

dB	Environmental Condition
0	Threshold of hearing
10	Rustle of leaves
20	Broadcasting studio
30	Bedroom at night
40	Library
50	Quiet office
60	Conversational speech (at 1m)
70	Average radio
74	Light traffic noise
90	Subway train
100	Symphony orchestra
110	Rock band
120	Aircraft takeoff
146	Threshold of pain

There are several sources of noise pollution that contribute to both indoor and outdoor noise pollution. Noise emanating from factories, vehicles, playing of loud speakers during various festivals can contribute to outdoor noise pollution while loudly played radio or music systems, and other electronic gadgets can contribute to indoor noise pollution. A study conducted by researchers from the New Delhi based National Physical Laboratory show that noise generated by firecrackers (presently available in the market) is much higher than the prescribed levels. The permitted noise level is 125 decibels, as per the Environment Protection (second amendment) Rules, 1999.

The differences between sound and noise is often subjective and a matter of personal opinion. There are however some very harmful effects caused by exposure to high sound levels. These effects can range in severity from being extremely annoying to being extremely painful and hazardous.

Effects of noise pollution on physical health

The most direct harmful effect of excessive noise is physical damage to the ear and the temporary or permanent hearing loss often called a temporary threshold shift (TTS). People suffering from this condition are unable to detect weak sounds. However, hearing ability is usually recovered within a month of exposure. In Maharashtra people living in close vicinity of Ganesh mandals that play blaring music for ten days of the Ganesh festival are usually known to suffer from this phenomenon. Permanent loss, usually called noise induced permanent threshold shift (NIPTS) represents a loss of hearing ability from which there is no recovery.

Below a sound level of 80 dB hearing loss does not occur at all. However, temporary effects are noticed at sound levels between 80 and 130 dB. About 50 per cent of the people exposed in this level.

Noise Control techniques

There are four fundamental ways in which noise can be controlled: Reduce noise at the source, block the path of noise, increase the path length and protect the recipient. In general, the best control method is to reduce noise levels at the source.

Source reduction can be done by effectively muffling vehicles and machinery to reduce the noise. In industrial areas noise reduction can be done by using rigid sealed enclosures around machinery lined with acoustic absorbing material. Isolating machines and their enclosures from the floor using special spring mounts or absorbent mounts and pads and using flexible couplings for interior pipelines also contribute to reducing noise pollution at the source.

However, one of the best methods of noise source reduction is regular and thorough maintenance of operating machinery. Noise levels at construction sites can be controlled using proper construction planning and scheduling techniques. Locating noisy air compressors and other equipment away from the site boundary along with creation of temporary barriers to physically block the noise can help contribute to reducing noise pollution. Most of the vehicular noise comes from movement of the vehicle tires on the pavement and wind resistance. However poorly maintained vehicles can add to the noise levels. Traffic volume and speed also have significant effects on the overall sound. For example doubling the speed increases the sound levels by about 9 dB and doubling the traffic volume (number of vehicles per hour) increases sound levels by about 3 dB. A smooth flow of traffic also causes less noise than does a stop-and-go traffic pattern. Proper highway planning and design are essential for controlling traffic noise. Establishing lower speed limits for highways that pass through residential areas, limiting traffic volume and providing alternative routes for truck traffic are effective noise control measures. The path of traffic noise can also be blocked by construction of vertical barriers alongside the highway. Planting of trees around houses can also act as effective noise barriers. In industries different types of absorptive material can be used to control interior noise. Highly absorptive interior finish material for walls, ceilings and floors can decrease indoor noise levels significantly. Sound levels drop significantly with increasing distance from the noise source. Increasing the path length between the source and the recipient offers a passive means of control. Municipal land-use ordinances pertaining to the location of airports make use of the attenuating effect of distance on sound levels. Use of earplugs and earmuffs can protect individuals effectively from excessive noise levels. Specially designed earmuffs can reduce the sound level reaching the eardrum by as much as 40 dB.

5. Thermal pollution

Sources: The discharge of warm water into a river or water body is usually called a thermal pollution. It occurs when an industry removes water from a source, uses the water for cooling purposes and then returns the heated water to its source.

Power plants heat water to convert it into steam, to drive the turbines that generate electricity. For efficient functioning of the steam turbines, the steam is condensed into water after it leaves the turbines. This condensation is done by taking water from a water body to absorb the heat. This heated water, which is atleast 15 °C higher than the normal is discharged back into the water body.

Effects: The warmer temperature decreases the solubility of oxygen and increases the metabolism of fish. This changes the ecological balance of the river. Within certain limits thermal additions can promote the growth of certain fish and the fish catch may be high in the vicinity of a power plant. However, sudden changes in temperature caused by periodic plant shutdowns both planned and unintentional can change result in death of these fish that are acclimatized to living in warmer waters.

Tropical marine animals are generally unable to withstand a temperature increase of 2 to 3 °C. Most sponges, mollusks and crustaceans are eliminated at temperatures above 37 °C. This results in a change in the diversity of fauna as only those species that can live in warmer water survive.

Control measures: Thermal pollution can be controlled by passing the heated water through a cooling pond or a cooling tower after it leaves the condenser. The heat is dissipated into the air and the water can then be discharged into the river or pumped back to the plant for reuse as cooling water. There are several ways in which thermal pollution can be reduced. One method is to construct a large shallow pond. Hot water is pumped into one end of the pond and cooler water is removed from the other end. The heat gets dissipated from the pond into the atmosphere.

A second method is to use a cooling tower. These structures take up less land area than the ponds. Here most of the heat transfer occurs through evaporation. Here warm waters coming from the condenser is sprayed downward over vertical sheets or baffles where the water flows in thin films. Cool air enters the tower through the water inlet that encircles the base of the tower and rises upwards causing evaporative cooling. A natural draft is maintained because of the density difference between the cool air outside and the warmer air inside the tower. The waste heat is dissipated into the atmosphere about 100 m above the base of the tower. The cooled water is collected at the floor of the tower and recycled back to the power plant condensers. The disadvantage in both these methods is however those large amounts of water are lost by evaporation.

Chapter - 12

Ecological Factors Affecting Crop Production

The factors are broadly classified into

1. INTERNAL FACTORS (Genetic or heredity factors)
2. EXTERNAL FACTORS (Environmental factors)

A. Climatic

B. Edaphic

C. Biotic

D. Physiographic

E. Socio-economic

1. Internal factors (Genetic factors)

The increase in crop yields and other desirable characters are related to Genetic make up of plants.

- » High yielding ability
- » Early maturity
- » Resistance to lodging
- » Drought, flood and salinity tolerance
- » Tolerance to insect pests and diseases

- » Chemical composition of grains (oil content, protein content)
- » Quality of grains (fineness, coarseness)
- » Quality of straw (sweetness, juiciness)

The above characters are less influenced by environmental factors since they are governed by genetic make-up of crop.

2. External factors

A. Climatic

B. Edaphic

C. Biotic

D. Phsiographic

E. Socio-economic

a. Climatic Factors

Nearly 50 % of yield is attributed to the influence of climatic factors. The following are the atmospheric weather variables which influences the crop production.

1. Precipitation

2. Temperature

3. Atmospheric humidity

4. Solar radiation

5. Wind velocity

6. Atmospheric gases

1. Precipitation

- » Precipitation includes all water which falls from atmosphere such as rainfall, snow, hail, fog and dew.
- » Rainfall one of the most important factor influences the vegetation of a place.

- » Total precipitation in amount and distribution greatly affects the choice of a cultivated species in a place.
- » In heavy and evenly distributed rainfall areas, crops like rice in plains and tea, coffee and rubber in Western Ghats are grown.
- » Low and uneven distribution of rainfall is common in dryland farming where drought resistance crops like pearl millet, sorghum and minor millets are grown.
- » In desert areas grasses and shrubs are common where hot desert climate exists
- » Though the rainfall has major influence on yield of crops, yields are not always directly proportional to the amount of Precipitation as excess above optimum reduces the yields
- » Distribution of rainfall is more important than total rainfall to have longer growing period especially in drylands

2. Temperature

- » Temperature is a measure of intensity of heat energy. The range of temperature for maximum growth of most of the agricultural plants is between 15 and 40 °C.
- » The temperature of a place is largely determined by its distance from the equator (latitude) and altitude.
- » It influences distribution of crop plants and vegetation.
- » Germination, growth and development of crops are highly influenced by temperature.
- » Affects leaf production, expansion and flowering.
- » Physical and chemical processes within the plants are governed by air temperature.
- » Diffusion rates of gases and liquids changes with temperature.
- » Solubility of different substances in plant is dependent on temperature.
- » The minimum, maximum (above which crop growth ceases) and optimum temperature of individual's plant is called as cardinal temperature

Crops	Minimum temperature °C	Optimum temperature °C	Maximum temperature °C
Rice	10	32	36-38
Wheat	4.5	20	30-32
Maize	8-10	20	40-43
Sorghum	12-13	25	40
Tobacco	12-14	29	35

3. Atmospheric Humidity (Relative Humidity - RH)

» Water is present in the atmosphere in the form of invisible water vapour, normally known as humidity. Relative humidity is ratio between the amount of moisture present in the air to the saturation capacity of the air at a particular temperature.

» If relative humidity is 100% it means that the entire space is filled with water and there is no soil evaporation and plant transpiration.

» Relative humidity influences the water requirement of crops

» Relative humidity of 40-60% is suitable for most of the crop plants.

» Very few crops can perform well when relative humidity is 80% and above.

» When relative humidity is high there is chance for the outbreak of pest and disease.

4. Solar radiation (without which life will not exist)

» From germination to harvest and even post harvest crops are affected by solar radiation.

» Biomass production by photosynthetic processes requires light.

» All physical process taking place in the soil, plant and environment are dependent on light

» Solar radiation controls distribution of temperature and there by distribution of crops in a region.

» Visible radiation is very important in photosynthetic mechanism of plants.

Photosynthetically Active Radiation (PAR - 0.4 – 0.7μ) is essential for production of carbohydrates and ultimately biomass.

0.4 to 0.5 μ - Blue – violet – Active

0.5 to 0.6 μ - Orange – red - Active

0.5 to 0.6 μ - Green –yellow – low active

» Photoperiodism is a response of plant to day length

Short day – Day length is 12 hours (Barley, oat, carrot and cabbage),

long day – Day length is > 12 hours (Barley, oat, carrot and cabbage),

day neutral – There is no or less influence on day length (Tomato and maize).

» Phototropism — Response of plants to light direction. Eg. Sunflower

» Photosensitive – Season bound varieties depends on quantity of light received

5. Wind velocity

» The basic function of wind is to carry moisture (precipitation) and heat.

» The moving wind not only supplies moisture and heat, also supplies fresh CO_2 for the photosynthesis.

» Wind movement for 4 – 6 km/hour is suitable for more crops.

» When wind speed is enormous then there is mechanical damage of the crops (i.e.) it removes leaves and twigs and damages crops like banana, sugarcane

» Wind dispersal of pollen and seeds is natural and necessary for certain crops.

» Causes soil erosion.

» Helps in cleaning produce to farmers.

» Increases evaporation.

» Spread of pest and diseases.

6. Atmospheric gases on plant growth

- CO_2 – 0.03%, O_2 - 20.95%, N_2 - 78.09%, Argon - 0.93%, Others - 0.02%.
- CO_2 is important for Photosynthesis, CO_2 taken by the plants by diffusion process from leaves through stomata
- CO_2 is returned to atmosphere during decomposition of organic materials, all farm wastes and by respiration
- O_2 is important for respiration of both plants and animals while it is released by plants during Photosynthesis
- Nitrogen is one of the important major plant nutrient, Atmospheric N is fixed in the soil by lightning, rainfall and N fixing microbes in pulses crops and available to plants
- Certain gases like SO_2, CO, CH_4, HF released to atmosphere are toxic to plants

B. Edaphic Factors (soil)

Plants grown in land completely depend on soil on which they grow. The soil factors that affect crop growth are

1. Soil moisture

2. Soil air

3. Soil temperature

4. Soil mineral matter

5. Soil organic matter

6. Soil organisms

7. Soil reactions

1. Soil moisture

- Water is a principal constituent of growing plant which it extracts from soil
- Water is essential for photosynthesis

» The moisture range between field capacity and permanent wilting point is available to plants.

» Available moisture will be more in clay soil than sandy soil

» Soil water helps in chemical and biological activities of soil including mineralization

» It influences the soil environment Eg. it moderates the soil temperature from extremes

» Nutrient availability and mobility increases with increase in soil moisture content.

2. Soil air

» Aeration of soil is absolutely essential for the absorption of water by roots

» Germination is inhibited in the absence of oxygen

» O_2 is required for respiration of roots and micro organisms.

» Soil air is essential for nutrient availability of the soil by breaking down insoluble mineral to soluble salts

» For proper decomposition of organic matter

» Potato, tobacco, cotton linseed, tea and legumes need higher O_2 in soil air

» Rice requires low level of O_2 and can tolerate water logged (absence of O_2) condition.

3. Soil temperature

» It affects the physical and chemical processes going on in the soil.

» It influences the rate of absorption of water and solutes (nutrients)

» It affects the germination of seeds and growth rate of underground portions of the crops like tapioca, sweet potato.

» Soil temperature controls the microbial activity and processes involved in the nutrient availability

» Cold soils are not conducive for rapid growth of most of agricultural crops

4. Soil mineral matter

» The mineral content of soil is derived from the weathering of rocks and minerals as particles of different sizes.

» These are the sources of plant nutrients eg; Ca, Mg, S, Mn, Fe, K etc.

5. Soil organic matter

» It supplies all the major, minor and micro nutrients to crops

» It improves the texture of the soil

» It increases the water holding capacity of the soil

» It is a source of food for most microorganisms

» Organic acids released during decomposition of organic matter enables mineralisation process thus releasing unavailable plant nutrients

6. Soil organisms

» The raw organic matter in the soil is decomposed by different micro organisms which in turn releases the plant nutrients

» Atmospheric nitrogen is fixed by microbes in the soil and is available to crop plants through symbiotic (Rhizobium) or non-symbiotic (Azospirillum) association

7. Soil reaction (pH)

» Soil reaction is the pH (hydrogen ion concentration) of the soil.

» Soil pH affects crop growth and neutral soils with pH 7.0 are best for growth of most of the crops

» Soils may be acidic (7.0)

» Soils with low pH is injurious to plants due high toxicity of Fe and Al.

» Low pH also interferes with availability of other plant nutrients.

C. Biotic Factors

Beneficial and harmful effects caused by other biological organism (plants and animals) on the crop plants.

1. Plants

- » Competitive and complementary nature among field crops when grown together
- » Competition between plants occurs when there is demand for nutrients, moisture and sunlight particularly when they are in short supply or when plants are closely spaced
- » When different crops of cereals and legumes are grown together, mutual benefit results in higher yield (synergistic effect)
- » Competition between weed and crop plants as parasites eg: Striga parasite weed on sugarcane crop

2. Animals

- » Soil fauna like protozoa, nematode, snails, and insects help in organic matter decomposition, while using organic matter for their living
- » Insects and nematodes cause damage to crop yield and considered as harmful organisms.
- » Honey bees and wasps help in cross pollination and increases yield and considered as beneficial organisms
- » Burrowing earthworm facilitates aeration and drainage of the soil as ingestion of organic and mineral matter by earthworm results in constant mixing of these materials in the soils.
- » Large animals cause damage to crop plants by grazing (cattle, goats etc)

D. Physiographic factors:

- » Topography is the nature of earth surface (leveled or sloppy) is known as topography. Topographic factors affect the crop growth indirectly.

» Altitude – increase in altitude cause a decrease in temperature and increase in precipitation and wind velocity (hills and plains)

» Steepness of slope: it results in run off of rain water and loss of nutrient rich top soil

» Exposure to light and wind: a mountain slope exposed to low intensity of light and strong dry winds may results in poor crop yields (coastal areas and interior pockets)

E. Socio-economic factors

» Society inclination to farming and members available for cultivation

» Appropriate choice of crops by human beings to satisfy the food and fodder requirement of farm household.

» Breeding varieties by human invention for increased yield or pest & disease resistance

» The economic condition of the farmers greatly decides the input/ resource mobilizing ability (marginal, small, medium and large farmers)

Chapter - 13

Ecology of Cropping System

Green revolution provided the much needed food security to the vast population of the Indian Nation but in spite of being one of the top producers of almost all agricultural commodities the productivity is very low when compared to international standards. The fatigue in the post-green revolution era is now a matter of serious concern. For achieving the required production targets to feed more than 1.5 billion in 2050, it is imperative to develop strategies that can sustain higher levels of production without an adverse affect on the environment. To maximize the production from the available resources and prevailing climatic conditions, need-based, location specific technology needs to be generated. Indian agriculture currently faces a host of diverse challenges and fresh constraints due to the ever growing population, increasing food and fodder needs, natural resource degradation, higher cost of inputs and concerns of climate change. A phenomenal increase in food grain production from 51 m.t. in 1950-51 to a record production of 304 m.t. of food grains in the year 2021-22 could be achieved using new as well as improved technologies involving high yielding varieties, increase in cropping intensity, enhanced use of fertilizers and other inputs. However, it is estimated that India would be requiring about 345 m.t. of food grains by 2030 (GOI, 2009), implying that we have to ensure an increase of about 5.5 m.t. food grains every year to achieve this target. On the other hand, the net cultivated area has hovered between 140 to 142 m.ha. since the last four decades with almost zero possibility of an increase in this area in future. It is also a fact that highly productive lands have been diverted from agriculture to infrastructural development, urbanization and other related activities. Under these circumstances, the only option is to increase the productivity vertically through ecological based cropping systems.

The term cropping system refers to the crops and crop sequences and the management techniques used on a particular field over a period of years. This term is not a new one, but it has been used more often in recent years in discussions about sustainability of our agricultural production systems.

In recent years a number of scientists have been studying the effects of cropping systems on the soil, water, and other natural resources in nearby fields where crops are grown and the general term agro-ecology or ecology of cropping systems has been coined to refer to this blend of ecology and agricultural sciences. Ecological services are means by which cropping systems can be shown to have positive effects on things like water quality, environment or soils.

A. Principles of agro-ecology

In a search to reinstate more ecological rationale into agricultural production, scientists and developers have disregarded a key point in the development of a more self-sufficient and sustaining agriculture: a deep understanding of the nature of agro-ecosystems and the principles by which they function. Given this limitation, agro-ecology has emerged as the discipline that provides the basic ecological principles for how to study, design and manage agro-ecosystems that are both productive and natural resource conserving, and that are also culturally sensitive, socially just and economically viable.

Agro-ecology goes beyond a one-dimensional view of agro-ecosystems - their genetics, agronomy, edaphology, and so on,- to embrace an understanding of ecological and social levels of co-evolution, structure and function. Instead of focusing on one particular component of the agro-ecosystem, agro-ecology emphasizes the interrelatedness of all agro-ecosystem components and the complex dynamics of ecological processes.

Agro-ecosystems are communities of plants and animals interacting with their physical and chemical environments that have been modified by people to produce food, fibre, fuel and other products for human consumption and processing. Agro-ecology is the holistic study of agro-ecosystems, including all environmental and human elements. It focuses on the form, dynamics and functions of their interrelationships and the processes in which they are involved. An area used for agricultural production, e.g. a field, is seen as a complex system in which ecological processes found under natural conditions also occur, e.g. nutrient cycling, predator/prey interactions, competition, and symbiosis and succession changes. Implicit in agro-ecological research is the idea that, by understanding these ecological relationships and processes, agro-ecosystems can be manipulated to improve production and to produce more sustainably, with fewer negative environmental or social impacts and fewer external inputs.

The design of such systems is based on the application of the following ecological principles.

1. Enhance recycling of biomass and optimizing nutrient availability and balancing nutrient flow.
2. Securing favorable soil conditions for plant growth, particularly by managing organic matter and enhancing soil biotic activity.
3. Minimizing losses due to flows of solar radiation, air and water by way of microclimate management, water harvesting and soil management through increased soil cover.
4. Species and genetic diversification of the agro-ecosystem in time and space.
5. Enhance beneficial biological interactions and synergisms among agro-biodiversity components thus resulting in the promotion of key ecological processes and services.

These principles can be applied by way of various techniques and strategies. Each of these will have different effects on productivity, stability and resiliency within the farm system, depending on the local opportunities, resource constraints and, in most cases, on the market. The ultimate goal of agro-ecological design is to integrate components so that overall biological efficiency is improved, biodiversity is preserved, and the agro-ecosystem productivity and its self-sustaining capacity is maintained. The goal is to design a quilt of agro-ecosystems within a landscape unit, each mimicking the structure and function of natural ecosystems.

B. Bio diversification of agro-ecosystems for sustainable cropping system

From a management perspective, the agro-ecological objective is to provide balanced environments, sustained yields, biologically mediated soil fertility and natural pest regulation through the design of diversified agro-ecosystems and the use of low-input technologies. Agro-ecologists are now recognizing that intercropping, agro-forestry and other diversification methods mimic natural ecological processes, and that the sustainability of complex agro-ecosystems lies in the ecological models they follow. By designing farming systems that mimic nature, optimal use can be made of sunlight, soil nutrients and rainfall.

Agro-ecological management must lead management to optimal recycling of nutrients and organic matter turnover, closed energy flows, water and soil conservation and balance pest-natural enemy populations. The strategy exploits the complementarities and synergisms that result from the various combinations of crops, tree and animals in spatial and temporal arrangements.

In essence, the optimal behaviour of agro-ecosystems depends on the level of interactions between the various biotic and a biotic components. By assembling a functional biodiversity it is possible to initiate synergisms which subsidize agro-ecosystem processes by providing ecological services such as the activation of soil biology, the recycling of nutrients, the enhancement of beneficial arthropods and antagonists, and so on. Today there is a diverse selection of practices and technologies available, and which vary in effectiveness as well as in strategic value. Key practices are those of a preventative nature and which act by reinforcing the "immunity" of the agro-ecosystem through a series of mechanisms. Various strategies to restore agricultural diversity in time and space include crop rotations, cover crops, intercropping, crop/livestock mixtures, and so on, which exhibit the following ecological features:

1. ***Crop rotations***: Temporal diversity incorporated into cropping systems, providing crop nutrients and breaking the life cycles of several insect pests, diseases, and weed life cycles.

2. ***Polycultures***: Complex cropping systems in which two or more crop species are planted within sufficient spatial proximity to result in competition or complementation, thus enhancing yields.

3. ***Agro-forestry systems***: An agricultural system where trees are grown together with annual crops and/or animals, resulting in enhanced complementary relations between components increasing multiple use of the agro-ecosystem.

4. ***Cover crops***: The use of pure or mixed stands of legumes or other annual plant species under fruit trees for the purpose of improving soil fertility, enhancing biological control of pests, and modifying the orchard microclimate.

5. **Livestock integration** in agro-ecosystems aids in achieving high biomass output and optimal recycling.

All of the above diversified forms of agro-ecosystems share in common the following features:

a. Maintain vegetative cover as an effective soil and water conserving measure, met through the use of no-till practices, mulch farming, and use of cover crops and other appropriate methods.

b. Provide a regular supply of organic matter through the addition of organic matter (manure, compost, and promotion of soil biotic activity).

c. Enhance nutrient recycling mechanisms through the use of livestock systems based on legumes, etc.

d. Promote pest regulation through enhanced activity of biological control agents achieved by introducing and/or conserving natural enemies and antagonists.

Research on diversified cropping systems underscores the great importance of diversity in an agricultural setting. Diversity is of value in agro-ecosystems for a variety of reasons:

» As diversity increases, so do opportunities for co-existence and beneficial interactions between species that can enhance agro-ecosystem sustainability.

» Greater diversity often allows better resource-use efficiency in an agro-ecosystem. There is better system-level adaptation to habitat heterogeneity, leading to complementarily in crop species needs, diversification of niches, overlap of species niches, and partitioning of resources.

» Ecosystems in which plant species are intermingled possess an associated resistance to herbivores as in diverse systems there is a greater abundance and diversity of natural enemies of pest insects keeping in check the populations of individual herbivore species.

» A diverse crop assemblage can create a diversity of microclimates within the cropping system that can be occupied by a range of noncrop organisms - including beneficial predators, parasites, pollinators, soil fauna and antagonists - that are of importance for the entire system.

» Diversity in the agricultural landscape can contribute to the conservation of biodiversity in surrounding natural ecosystems.

» Diversity in the soil performs a variety of ecological services such as nutrient recycling and detoxification of noxious chemicals and regulation of plant growth.

» Diversity reduces risk for farmers, especially in marginal areas with more unpredictable environmental conditions. If one crop does not do well, income from others can compensate.

C. Importance of sustainable agro-ecosystems

Most people involved in the promotion of sustainable agriculture aim at creating a form of agriculture that maintains productivity in the long term by:

» Optimizing the use of locally available resources by combining the different components of the farm system, i.e. plants, animals, soil, water, climate and people, so that they complement each other and have the greatest possible synergetic effects;

» Reducing the use of off-farm, external and non-renewable inputs with the greatest potential to damage the environment or harm the health of farmers and consumers, and a more targeted use of the remaining inputs used with a view to minimizing variable costs;

» Relying mainly on resources within the agro-ecosystem by replacing external inputs with nutrient cycling, better conservation, and an expanded use of local resources;

» Improving the match between cropping patterns and the productive potential and environmental constraints of climate and landscape to ensure long-term sustainability of current production levels;

» Working to value and conserve biological diversity, both in the wild and in domesticated landscapes, and making optimal use of the biological and genetic potential of plant and animal species; and

» Taking full advantage of local knowledge and practices, including innovative approaches not yet fully understood by scientists although widely adopted by farmers.

Agro-ecology provides the knowledge and methodology necessary for developing an agriculture that is on the on e hand environmentally sound and on the other hand highly productive, socially equitable and economically viable. Through the application of agro-ecological principles, the basic challenge for sustainable agriculture to make better use of internal resources can be easily achieved by minimizing the external inputs used, and preferably by regenerating internal resources more effectively through diversification strategies that enhance synergisms among key components of the agro-ecosystem.

The ultimate goal of agro-ecological design is to integrate components so that overall biological efficiency is improved, biodiversity is preserved, and the agro-ecosystem productivity and its self-regulating capacity is maintained. The goal is to design an agro-ecosystem that mimics the structure and function of local natural ecosystems; that is, a system with high species diversity and a biologically active soil, one that promotes natural pest control, nutrient recycling and high soil cover to prevent resource loss.

Ecology of cereal based cropping systems aims at

» use natural resources efficiently

» provide stable and high returns

» do not damage the environment

Ecology of intercropping systems: For successful intercropping, certain important requirements are

» The time of peak nutrient demands of component crops should not overleap (eg. in maize + green gram intercropping system, the peak nutrient demand period for green gram is around 35 DAS while it is 50 days for maize.)

» Competition for light should be minimum among the component crops

» Complementary should exist between the component crops

» The differences in maturity of component crops should be at least 30 days

Intercropping systems provide:

1. Insurance against total crop failure under aberrant weather conditions or pest epidemics

2. Increase in total productivity per unit land area

3. Judicious utilization of resources such as land labour and inputs

Inter cropping in cereals: Inter-cropping with cereal is an excellent way of improving the resource utilization because the serial utilizes the rainy season resources while late maturing crops exploits the post-rainy season resources such as residual moisture.

Sorghum is most commonly inter-cropped with pigeon pea on a variety of soils. Sorghum is harvested after 3 ½ to 4 ½ months and pigeon pea matures in about 6-9 months depending on the genotype.

In a mono-cropping of groundnut resources like rainfall, temperature and solar radiation are utilized only rainy season, however when red gram is introduced as an inter-crop, these resources are used upto the end of winter season and also benefit of late shower rains.

Soybean is also a good compatible companion crop with maize for maintain soil fertility.

Pearl millet is a quick tillering and fast growing crop that attains full canopy development within 20-30 days of seedling establishment. It can be inter-cropped with groundnut, black-gram or castor.

Inter cropping in pulses: For pigeon pea, short duration grain legumes such as black-gram, grrengram and soybean are the best companion crops in peninsular India. Groundnut is also a suitable inter-crop commonly grown with pigeon pea in South India.

Inter-cropping in cotton: It is initially a slow growing crop. Any short duration and fast growing crops such as groundnut, black-gram, green-gram or cluster bean are the compatible companion crops.

Inter cropping in sugarcane: Sugarcane is slow growing up to 80-90 days. Since the crop is planted in rows 0.8-1.0 m apart, considerable space is available for inter-cropping. Short duration crops maturing in 80-90 days can be advocated as inter-crops. Black-gram and soybean are found suitable. The green manure, dhaincha can be sown in the inter-space and incorporated at about 2 months.

D. Ecology of legume based cropping systems

The primary question today is how to produce crops in a manner that is

- » Economically profitable
- » Environmentally acceptable and sustainable
- » Legumes offer an alternative source of nitrogen
- » Usually aid in soil erosion control
- » Improved tillage practices and
- » Offer potential for improvement of surface soil

i. Legume based cropping systems could help to

- Increase crop productivity and soil organic matter levels thereby enhancing soil quality as well as having the additional benefit of sequestering atmospheric Carbon
- Increased symbiotic nitrogen fixation capacity
- Increased phosphorus recovery from the soil
- Improving organic carbon level of soil
- Grain legume yield limitations
- Cropping systems to take advantage of the
- Multi-dimensional benefits of grain legumes
- Legumes rotations have long-term benefit on soil structure compared to continuous grain cropping; result in enhanced soil organic content and mineralizable N
- This provides better control of N availability
- Improved soil structure, less energy for cultivation, and less erosion
- Reduction in erosion rate, over a period of decades, can have a major influence on the properties and productivity of some soils

ii. Complimentary benefits of legume cropping systems

- Legumes can be used as a green manure
- Green manuring with legumes involves growing the plants, then slashing the crop and leaving it on the soil surface. Leaving the crop on the soil surface has additional benefits, as it also reduces soil erosion and conserves soil moisture
- Leguminous shrubs can also improve soil quality. One option is to grow the shrubs in rotation with the cereal crop, cutting the shrubs to produce mulch and fodder. However, this alternative does involve the loss of a growing season
- A second option is to grow leguminous shrubs as hedge rows either within or

around the field. When the bushes are pruned, the clippings can be applied as mulch to the soil surface

- An added advantage of these shrubs is that farmers can also produce honey during the flowering period added advantage of these shrubs is that farmers can also produce honey during the flowering period

iii. Effect of legumes on soil organic carbon

- In agricultural systems, optimization of carbon and nitrogen cycling through soil organic matter can improve soil fertility and yields
- while reducing negative environmental impact that the use of low carbon-to-nitrogen organic residues to maintain soil fertility, combined with greater temporal diversity in cropping sequences and
- Significantly increases the retention of soil carbon and nitrogen and environmental quality

E. Ecological sustainability through crop rotations

- Improved yields and crop quality
- Enhanced erosion protection
- Reduced runoff and pollutants in runoff
- Increased soil organic matter
- Increased biological activity in the soil
- Improved efficiency of nutrient cycling and utilization
- Reduced soil compaction and improved soil condition
- Increased diversity and wildlife habitat
- Fixation of nitrogen for subsequent crops
- Minimized potential of pesticide resistant strains among pest ppulations
- Aid in control of crop pests
- Can facilitate development and implementation of a compatible nutrient management lan

» Improved water use efficiency

i. Impacts of sustainable crop rotations on soil quality

» High residue crops and perennials (sod) increase SOM

» High residue crops and perennials (sod) reduce soil erosion

» Diversity in crop rotations increases biological activity in the soil and diversity of soil biota

» Many perennial species and some annuals have deep extensive root systems that reduce soil compaction and improve soil condition

» High residue crops and perennials reduce the amount of sediment and other particulate matter leaving a field

» Live vegetation captures and cycles excessive nitrates and other nutrients in the soil

» Crop residues and perennial vegetation help reduce and filter runoff

ii. Characteristics of sustainable crop rotations

» Include a diversity of crops/vegetative types

» Use a combination of species (crops and covers) and a sequence to minimize pest problems, break pest cycles, and promote healthier crops

» Maximize benefits and sustainability by keeping live cover on the land to the extent feasible

» Provide for nutrient needs of all crops, including covers

» Include deep rooted grass cover crops, as needed, to capture and cycle nitrates in the soil and to reduce soil compaction

iii. Using cover crops to enhance crop rotations

» Can provide an opportunity to overcome weaknesses in crop rotations without sacrificing profitability or the production system

» Crop rotations & cover crops are keys to sustainable cropping systems

F. Benefits of cover crops in crop rotations

- » Increase soil cover and soil organic matter
- » Leguminous cover crops can produce nitrogen for subsequent crops
- » Attract beneficial insects and other predators of insect pests
- » Can provide weed suppression and allelepathic effects
- » can serve as a trap crop for control of insect pests
- » Can reduce soil compaction (e.g. forage radishes)
- » Can capture and cycle residual nitrogen or nitrogen released from decaying residues (cereal rye and forage radishes)
- » Some species provide a bio-fumigation effect e.g. certain brassicas.
- » Enhance biological activity in the root zone

i. Contributions of cover crops to weed management

- » Allelo-pathic effects
- » Suppression by residue cover
    - Increases with increasing quantity
    - Declines with decomposition of residue
    - Provides greatest control of small-seeded species that require light for germination.
- » Maximize effects by use of cover crops that produce large amounts of biomass and use of implements that pack or compress mulch.

Sum up

In a world of increasing demands for food and plant products, at the same time of greater economic and environmental pressures on land and resources, it is apparent that adopting ecological principles need to become a major components of future cropping systems which address:

- » Balancing productivity, profitability, and environmental health is a key challenge for agricultural sustainability.

» Most crop production systems in India are characterized by low species and management diversity, indiscriminate use of agrichemicals, and large negative impacts on the environment.

» We hypothesized that cropping system diversification would promote ecosystem services that would supplement, and eventually displace, synthetic external inputs used to maintain crop productivity.

Chapter - 14

Response of Crops to Light, Temperature and CO_2 on Growth and Development

A. Solar radiation and crop response

Solar radiation is the main source of energy for photosynthesis. This energy is available for plant growth mainly when plants absorbs and utilized by the crop canopy. Not all the radiation is useful for photosynthesis; only 400-700 nm is used. This fraction is considered as photosynthetically active radiation (PAR). Plants do respond differently to radiation both in visible and infrared rays (IR). In visible range, PAR is strongly absorbed by the pigment. IR range contribute 700-1300 nm is mainly used for water absorption.

Absorbed PAR (APAR) is algebraic sum of incoming and outgoing flux densities measured above and below plant canopy. Four independent PAR flux densities are measured, two above the plant canopy and two below plant canopy, required to access APAR.

Above plant canopy measurements are

1. Incident PAR flux density i.e., I_O
2. PAR flux density from the canopy i.e., Rc

Below crop canopy measurements required are

1. PAR flux density transmitted through the canopy to the soil surface (T_C)
2. PAR flux density reflected by soil surface (Ro)

$$APAR = (I_O + R_O) - (T_C + R_C)$$

Instantaneous APAR measurement are usually expressed in photo unit i.e., micromoles of energy/m^2/sec or in terms of energy unit watts/m^2.

The term intercepted radiation and absorbed radiation are often used inter changeably in literature. There is great variation and important distinction between these 2 terms. Because all intercepted radiation is not necessarily be absorbed although photons may be intercepted but not all intercepted absorbed but some are scattered, absorbed, reflected and transmitted. In PAR range scattering by green leaves is low to extent of 15 per cent and intercepted PAR is therefore difference in the flux density above and below the plant canopy. The 2 measurements required are the incoming PAR flux density and PAR flux density transmitted through plant canopy to soil surface IPAR = $I_O - I_C$.

Russel *et al.* (1989) showed that as long as canopy is complete and consists strictly green leaves IPAR is almost equal to APAR because of healthy green leaves which reflect very little PAR radiation but for open canopies, canopies with green or brown background IPAR and APAR will differ a lot.

Dry Matter Production in any crop or cropping system is often proportional to quantity of light absorbed by crop canopy. This is because photosynthetic rate of individual leaves responds linearly to increase in radiation levels upto the level when they become saturated, while canopy photosynthesis will also show similar response but saturation occur at higher level of incoming radiation. Because leaf angle mutual shading reduces the actual irradiation of individual leaf considering strong relation between light intensity and DMP. The DMP is almost proportional to mean photosynthetic rate. Therefore, when water, nutrient and temperature are not limiting, the quantity of DM produced by crop is directly proportional to S fet.

W (g/m^2) = S fet.

S= Daily mean solor radiation (MS/m^2/day)

f= seasonal mean fractional interception of radiation by canopy

e= seasonal mean conversion coefficient

t=canopy duration in days

Conversion coefficient (e) is quantity of DM produced/unit of light intercepted and also referred to as light use efficiency or radiation use efficiency. However, use efficiency in this context is not appropriate.

While 'K' the extinction co-efficient depends on angle of leaf and distribution of leaf on stem. With increase in light or water or nutrient biomass production can be increased either by increased greater resource capture of light i.e., either due to higher 'F' values or due to 'K' values or by increased conversion co-efficient (high 'e' value). The same equation will hold good for components in cropping system but incident radiation, remain unchanged as a whole irradiation received by shorter component may be reduced by shading in cropping system with component of cropping system is strongly fallen on the other component.

In cropping system like pearl millet-pigeonpea, sunflower-pigeon pea, the intercepted PAR is more compared to sole crop under rainfed condition thus always intercrop crop will have more LUE (Shivaramu and Shivashankar, 1994).

i. Reflection, transmission and absorption of light

Maximum reflection transmission is seen in green light and IR range. The impression of green colour of plants depends on high reflectivity of solar radiation and greater sensitivity of human eye for green colour. These aspects of solar radiation in crop canopy are biologically significant.

1. Intensity of radiation i.e., amount of radiant energy falling on unit surface area on unit time.
2. Spatial distribution in time is input for photo periodic response
3. Radiation distribution in time is output for photo periodic response.

The rate of photosynthesis depends on availability of photosynthetically active radiation intercepted by leaf to rate of transpiration from plant canopy which also controlled to greater extent by radiation energy.

ii. Factors influencing radiation distribution with in plant canopy

1. Transmisivity of leaves
2. Leaf arrangement and inclination
3. Canopy architecture
4. Plant density
5. Plant height
6. Angle of the sun

1. **Transmisivity of leaves:** Herbs and grasses including cereals have transmisivity of 5-10 per cent where as broad leaves of evergreen plants have little lower transmisivity of 2-8 per cent. This transmisivity will vary with age. Growth of young leaf is relatively high and with maturity it declines and rises as the leaf turns yellow. Growth of leaf is directly related to chlorophyll content. Log of T decrease with increase in the chlorophyll content.

2. **Leaf arrangement and inclination**: If the leaves are erect 10 per cent of radiation is horizontally displayed in continuous layer and only 15 per cent of mostly green region condition penetrate to second region. However, leaves are rarely arranged horizontally on stem. Interception of horizontal leaf that are erect will be in the ratio of 1:0.44. It has been found that the total leaf area of plant equal to ground area, then transmisivity will be 74.5 per cent for more upright leaves and 50 per cent of somewhat erect leaves. In full sunlight optimum leaf inclination for efficient light use is 81° angle. A leaf placed at optimum inclination will have 4.5 per cent higher efficiency than the horizontal leaves.

3. **Plant density:** With plant canopy is dense there will be more light interception when compared to sparse canopy. Therefore, intercropping systems will have added advantage of utilizing PAR more efficiently than sole crops.

4. **Plant height:** Plant height increase interception of light by canopy.

5. **Angle of sun:** Per cent of light interception by plant canopy is minimum at noon whereas, maximum at morning and evening hours. Over last 2 decades there has been increase in radiation measurement of solar radiation on crop canopy and its use in judging plant productivity. The growth of crop is often increasingly analyzed based on amount of intercepted solar radiation and efficiency of its conversion into dry matter production.

Srinivas *et.al.,* (1995) opined that APAR is significantly lower under sole pigeon pea whereas, it was higher with sole pearl millet at 50 day after sowing. However, during later period pearl millet + pigeon pea or pearl millet +sunflower intercropping system PAR absorbed is more when compared to sole crop and after harvest of pearl millet and sunflower, sole pigeon pea absorbed more PAR compared to pearl millet or sunflower. The relationship between cumulative APAR and DMP showed that more efficient conversion of APAR was found in pearl millet + pigeon pea intercropping system either 2:1 or 4:1 row proportion.

Extinction co-efficient (K): EC is angle of leaf. In different climates 'K' value is generally influenced by row spacing, plant population and plant geometry. K value is more under narrow row geometry than wider row geometry because of more competition. This enables more interception of PAR under close spacing, on other hand, 'K' values decrease with increase in row spacing (Shinde *et al.* 1996).

6. Leaf area index

It is a yield determining factor in most of field crops because it influences largely both light interception as well as transpiration ratio i.e., amount of light intercepted by leaf and amount of water lost their transpiration. Presently leaf area meters are available which can determine the leaf area accurately and rapidly. It can also be estimated by using formula based on leaf length and maximum width and correction factor K. K is used because leaf is not exactly rectangle.

Bollero *et al.* (1996) and Yashida (1976) used 'K' factor for estimation of leaf area in rice and maize and it is 0.75 for all stages except seedling stage and maturity. For rice K value is lower at seedling stage and it is 0.67. Canopy light interception and photosynthesis are closely related to critical values of LAI. These critical values are required to intercept 95 per cent of intercepted radiation and values will increase until maximum value or attained around flowering and then starts declaiming. Maximum observed values are

>5.5 for rice, maize, potato and sugar beet

4.5-5.0 – soybean, common bean

3.5 – spring wheat

Leaf area is greatly influenced by climate factors.

B. Carbon dioxide and crop response

Elevated atmospheric CO_2 concentration and associated climate changes may affect crop production as well as world food supply in coming decades. During later part of 21st century crops are expected to be grown in environment with twice the present atmospheric concentration with average temperature being 2.5 °C warmer than present temperature. In addition to change in climatic conditions unexpected late springs and early frost and periodic episodes of heat and drought are predicted to occur more frequently due to global warming. These changes could influence climate effect on many aspects of crop growth and development thus it may decrease crop yield as well as quality. Current knowledge about interaction of CO_2 with key environmental factor is insufficient to show definite conclusions regarding magnitude of its effect even the direction of potential future changes on crop yield cannot be predicted.

Natural gases which cause global warming include CO_2, CO, nitrous oxide, nitric oxide, nitrogen dioxide (NO_2) methane, ozone. In addition artificial gasses posses the attributes similar to natural gases have which been introduced into atmosphere since 1930.

Chlorofluorocarbons and Chlorofluorohydrocarbons are collectively called as CFC's. These also trap solar radiation with in earth atmosphere and also behave similar to those in greenhouses therefore they termed as greenhouse graces or active radiative gases. Global warming and its potential effect on increasing temperature and rising sea level are worldwide concern. It has been estimated that temperature may increase 1.1-1.9 °C sea level rise between 0.14 to 0.24 meters by 2020.

Increase in concentration of greenhouse gases in atmosphere is predicted to rise mean temperature by 2.3°C by 2050. Together with more frequent episodes of more frequent water deficit. Atmospheric CO_2 in not only important from point of plant growth but also important for global energy balance. Because, CO_2 is most important greenhouse gases and increase in its concentration above 50 per cent causes radiative forcing. Concentration of atmospheric CO_2 have raised from about 280 μ mole/mole of air to 362 μ mole/mole in past nine decades. It is predicted to continue with an average of 15 μ mole/mole/year. During last 200 years, there has been increase in concentration of CO_2 to extent of 30 per cent. If developed countries continue their reluctance in reducing use of fossil fuels. It is inevitable that CO_2 concentration will reach between 510 to 760 μ mole/mol in next 50 years. An equation for predicting increased concentration of CO_2 as function of year has been developed by Gouduan (1995). CO_2 μ mole/mol. = 285 + 52 exp (0.024 (t – 1980). This equation fit well in prediction of CO_2 in future. Formula assume pre-industrial level of CO_2 i.e., 285 μ mole/mol CO_2 level of 337 for the reference year of 1980.

i. Green house effects (GHC)

Two major greenhouse substances are water vapour and water ice cloud. These are collectively responsible for >90% of GHE that keep the earth about 33 °C warmer than it would have been otherwise in the absence of water vapour and cloud. These substances have very short life time and quantities of water vapour adjust very quickly to long term climatic effect which is represented by changes in so called greenhouse gases.

Methane has increased at an average of 0.8% per year in the last 20 years and it is provided by the process such as

1. Gas and coal industrial combustion
2. Submerged rice fields
3. Due to cattle yards
4. Natural wet and or flooded lands.

Methane is also destroyed by chemical reaction caused by UV sunlight involving very active hydroxyl radicals. Methane was increased at rate of 1.2 % during 1970's but now at the rate less than 0.3%. Global methane emissions are decreased but methane destruction rates are either increasing.

N_2O has increased at rate of 0.2-0.3% per year and like methane it has wide range of poorly understood natural as well as anthropogenic regional sources. It is also destroyed by O_3 layer as that of CFC's

The CFC's are increased at the rate of 5 per cent per year during 1980's and now they are increasing at the rate of <3% as a result of Montreal protocol. It is also important to note that potency of both NO and CFC's offset significantly by stratospheric ozone which is destroyed by these gases. Although it is responsible for only small percentage of direct greenhouse warming, they have indirect influence on plant growth. Ex: Increase in temperature due to these gases leads to increased water vapour content and cloud content in atmosphere which accelerate because warming of atmosphere.

This water related processes are very important but not quantitatively well understood. Therefore, it leads to much uncertainty in current climate models. Any climate changes associated with increase in CO_2 will have potential effect on crop physiology, growth and grain yield of crops, because warming of atmosphere alters the

1) rate of photosynthesis
2) photorespiration
3) plant biosynthesis and accumulation
4) phenological development of crops etc.

So increased temperatures coupled with either changes in precipitation pattern will modify date of planting in several crops and results in distribution of crops.

ii. Soil carbon and its influence on climate change

Future changes in climate, atmosphere CO_2 concentration and land use will decisively influence the fate of carbon stored in soil thereby affecting CO_2 concentration in the atmosphere as well as climatic condition of the earth. Climate and soil properties are key factors in determining production as well as decomposition process of plant litter. In general, soil organic carbon is a major source of CO_2 that respond to changes in climate and atmosphere CO_2 concentration. Therefore, knowledge of C estimates in different soils is very important for managing and manipulating atmosphere CO_2 concentration although, soil and vegetation seen too small compared to ocean, it is potentially more stable in short time. The C balance of terrestrial ecosystem can be changed markedly by direct activity of humus such as deforestation, burning of fossil fuels, indiscriminate use of chemicals and fertilizers land use pattern that release trace gases which enhances GHE.

iii. CO_2 and plant growth

The current global rise in atmosphere CO_2 concentration has stimulated extensive research on response of agricultural crops to elevated CO_2 and its effect on plant growth. Plant growth stimulation response to elevated CO_2 will vary with photosynthetic pathway of crops. In addition, it also depends on growth stage, nutrient availability and plant species. In general, C_3 plants are more responsive to photosynthesis CO_2 level compared to C_4 plants. Increase in plant growth due to increased CO_2 concentration is generally due to high assimilation rate/unit leaf area (but not always). According to Grotenluis and Bugbee (1997) the effect of near optimal CO_2 (1200 μ mole/mol) and supra optimal CO_2 (2400 μ mole/mol) level on yield of a cultivar of hydroponically grown wheat showed that. vegetative growth was increased by 25% and seed yield by 15% in both varieties at near optimum CO_2 concentration and yield increase was primarily due to increase in number of heads/m^2. However, increase in CO_2 concentration (2400 μ mole/mol) decreased seed yield by 12% in one cultivar and 15% in the other. This is because of toxic effect of CO_2 are similar to light level ranges as well as light saturation capacity. Increased CO_2 affects plant growth mainly through leaf conductance, water vapour and CO_2 exchange. It is observed that leaf conductance in C_3 annual crop decreased by 34% when CO_2 concentration was twice the atmospheric level.

Allen *et al.* (1991) reported that rice grain yield increased with increase in CO_2 concentration and increased grain yield was associated generally with increase in tillering and increase in number of panicles (but not always) and yield of rice cultivar IR-30 was declined by 10% for each 10 °C increase day and night temperature above

28 and 21 °C, respectively in presence of elevated CO_2. Elevated CO_2 will have little effects in ameliorating the high temperature response. Sharp decrease in number of filled grains/panicle has caused reduction of grain yields in high temperature. Plant responses to CO_2 also depend on stage of crop development. Total dray matter production in spring wheat was increased only when young plants are exposed to photosynthesis CO_2 prior to floral initiation.

As per Neals and Nicholus (1978), 10 day old wheat plants responded to a increased CO_2 with greater growth and higher assimilation rates but, 24 day old plants have reverse response. Growth and development of barley also depended on plant growth stage. Slight response was observed in young plants. Photorespiration in C_3 plants decreases with increase in CO_2 concentration. Therefore, the rate of ratios of Rubp oxygenase to Rubp carboxylation reduces. At 20 °C, increase in CO_2 from 350 to 1200 μ mole/mol that decreased the rate of photorespiration in wheat from 24.6% to 20.2%.

Idro and Idro (1994) reported that relative growth enhancing effect of CO_2 are greatest when the other resources are limiting or plants are grown in suboptimal environment including those contaminated by air pollutants such as O_3. Photosynthesis CO_2 concentration reduce stomatal conductance, transpiration rates and improve water use efficiency, stimulate high rate of photosynthesis and increased light use efficiency.

Accumulation of photosynthates during long term exposure to elevated CO_2 may reduce the activity of key enzyme in the photosynthesis synthetic C-reduction cycle which will increase nutrient use efficiency. This effect will have major consequence in agriculture and natural ecosystem due to rise in atmospheric CO_2 and climate changes.

iv. CO_2 and photosynthesis

It is generally understand that photosynthesis CO_2 levels stimulate photosynthesis in crop plants. In a survey of 60 experiments Drake *et al.* (1991) reported that plant growth with elevated CO_2 concentration was increased by 58% when compared to rates with normal ambient concentration of CO_2. In addition to this, CO_2 has potential to regulate the reactions within the photosynthesis system. Ex: Binding Mn on the donor's site of photo system I and disrupting quinine binding sites on the acceptor sites of photo system II and then activating 1.5-ribulose biphosphate carboxylase and oxygenase. All these processes have exhibit high affinity for CO_2 and they become saturated at ambient CO_2 concentration whereas Rubisco has low affinity for CO_2 carboxylation reaction. This reaction will not be saturated at ambient CO_2 concentration.

Baker *et al.* (1990) reported that photosynthetic rate increased in rice with increase in CO_2 from 160 to 500 μ mole/mol followed by leveling off at super ambient concentration of 600-900 μ mole/mol.

In predicting response of photosynthesis by C_3 plants to increased CO_2 concentration, it has been recognized that photosynthetic stimulation from increased CO_2 usually increases extensively with increased temperature. This temperature dependence has firm theoretical basis related to temperature dependent aqueous O_2 to CO_2 solubility and to kinetic characteristics of Rubisco.

Bounce (1998) reported that short term stimulation of photosynthesis existed for wheat and barley with doubling CO_2 concentration from 350 to 700 μmol/mol. However, photosynthesis decreased when temperature was lowered from 30-10 ^{0}C at high photon flux.

v. CO_2 and water use efficiency(WUE)

WUE is the ratio of photosynthetic CO_2 assimilation to transpiration/unit leaf area. Increased CO_2 generally reduces stomatal conductance. Decrease in leaf conductance and increase in photosynthetic rate at increased CO_2 concentration results in high WUE for C_3 plant. In temperate plants, increased CO_2 concentration reduces water loss primarily through stomatal closure. In tropical plants increased WUE is attributed to combined effect of reduced stomatal conductance along with increased photosynthetic activity. It has been observed for both C_3 and C_4 plant in tropical species.

vi. CO_2 and radiation use efficiency (RUE)

RUE is defined as ratio of DMP/unit of intercepted light g/J and RUE was improved with increase in CO_2 concentration. Gallo *et al.* (1993) reported RUE is often critical aspect in crop growth model to relate dry matter production and energy received by crop canopy.

Plink *et al.* (1998) observed that cotton crop grown under free air C enrichment at 55 Pa has highly significant increase in RUE from 20 to 22% in consecutive years regardless of whether crops are grown with full irrigation or only with 50% optimum water supply.

vii. Management strategies for CO_2

Increasing CO_2 concentration in the atmosphere is indisputable. The IPCC 1990 predicted atmosphere CO_2 will continue to increase by 700 µmol/mol by end of 21 century mainly due to combustion of fossil fuel and deforestation. Results from climate model indicated that increased absorption of long wave radiation from increased concentration of CO_2 and other green house gases like methane, N_2O could alter climate globally. Therefore, managing terrestrial ecosystem especially forests to capture and store a atmospheric C has been proposed for reducing rate of increase atmosphere CO_2. Ex: planting of forest trees on marginal land provide many opportunity for C sequestration.

Cultivation generally decreases soil organic C content. No tillage system is proposed as an alternate for reducing soil degradation. No tillage leads to increased carbon in top 5-10 cm profile in the soil when compared to conventionally tilled soil. Other agricultural practices such as management of agricultural soils and range lands improved efficiency of fertilizer use; restoration of degraded agriculture lands can reduce emission of CO_2 to atmosphere.

According to Curtin *et al.* (1998), agricultural land if properly managed can sequester 50 to 75 per cent of CO_2 emitted from agriculture activity in next 30 years in countries like Canada. He also suggested management option to enhance C storage in Canada soils which include

(a) decreasing summer fallow

(b) reducing tillage operation

(c) including legume in crop rotation

(d) decreasing summer fallow

(e) incorporating residue of grasses on marginal land.

C sequestration is also important in considering effect of farming practices on greenhouses gases emission. Cropping system and practices that enhance C sequestration on atmosphere CO_2 are beneficial. Increase in soil O.C. with N fertilizer contributes to improved soil quality, productivity and improved efficiency of C sequestration in soil.

C sequestration can also be enhanced by increasing crop residue from adequate N fertilizer and by increasing cropping intensity and adequate fertilization can contribute positively to mitigate agricultural effects on atmosphere CO_2 level and their effect on global climate change.

C. Temperature and crop response

Temperature is a primary factor affecting the rate of plant development. Warmer temperatures expected with climate change and the potential for more extreme temperature events will impact crop production. Pollination is one of the most sensitive phenological stages to temperature extremes across all species and during this developmental stage temperature extremes would greatly affect production. Few adaptation strategies are available to cope with temperature extremes at this developmental stage other than to select for plants which shed pollen during the cooler periods of the day or are indeterminate so flowering occurs over a longer period of the growing season. In controlled environment studies, warm temperatures increased the rate of phenological development. However, there was no effect on leaf area or vegetative biomass compared to normal temperatures. The major impact of warmer temperatures was during the reproductive stage of development and in all cases grain yield in maize was significantly reduced by as much as 80–90% from a normal temperature regime. Temperature effects are increased by water deficits and excess soil water demonstrating that understanding the interaction of temperature and water will be needed to develop more effective adaptation strategies to offset the impacts of greater temperature extreme events associated with a changing climate.

i. Crops response to temperature

Rate of plant growth and development is dependent upon the temperature surrounding the plant and each species has a specific temperature range represented by a minimum, maximum, and optimu for a number of different species typical of grain, fodder or fruit production. The expected changes in temperature over the next 30–50 years are predicted to be in the range of 2–3 °C Intergovernmental Panel Climate Change (IPCC-2007). Heat waves or extreme temperature events are projected to become more intense, more frequent, and last longer than what is being currently been observed in recent years. Extreme temperature events may have short-term durations of a few days with temperature increases of over 5 °C above the normal temperatures. Extreme events occurring during the summer period would have the most dramatic impact on plant productivity. Recent studies on the effect of temperature extremes, frost and eat, in wheat (*Triticum aestivum* L.) revealed that frost caused sterility and abortion of formed grains while excessive heat caused reduction in grain number and reduced duration of the grain-filling period. Daily minimum temperatures will increase more rapidly than daily maximum temperatures leading to the increase in the daily mean temperatures and a greater likelihood of extreme events and these changes could have detrimental effects on grain yield. If these changes in temperature are expected to occur over the next 30 years then understanding the potential impacts on plant growth and development will help develop adaptation strategies to offset these impacts.

Responses to temperature differ among crop species throughout their life cycle and are primarily the phenological responses, i.e., stages of plant development. For each species, a defined range of maximum and minimum temperatures form the boundaries of observable growth. Vegetative development (node and leaf appearance rate) increases as temperatures rise to the species optimum level. For most plant species, vegetative development usually has a higher optimum temperature than for reproductive development. If we depict the range of temperatures in the following diagram (Fig.25) then the definition of extreme temperatures affecting plant response will be species dependent. For example, an extreme event for maize (*Zea mays* L.) will be warmer than for a cool season vegetable (broccoli, *Brassica oleracea* L.) where the maximum temperature for growth is 25 °C compared to 38 °C. In understanding extreme events and their impact on plants we will have to consider the plant temperature response relative to the meteorological temperature (Jerry Hatfield, 2015).

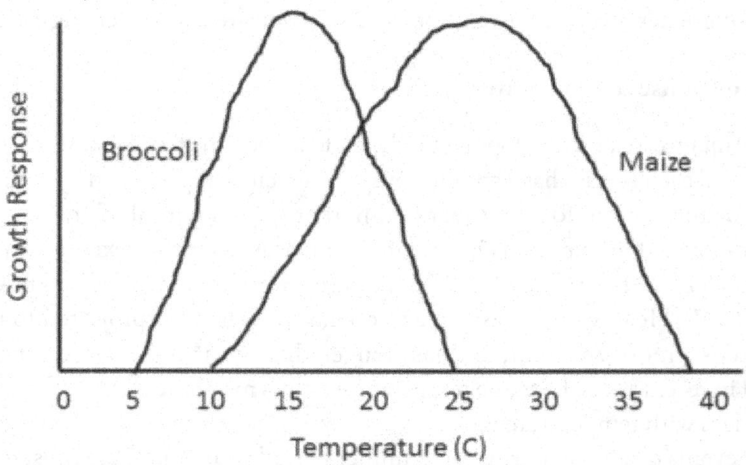

Fig.25: Temperature response for maize and broccoli plants showing the lower, upper and optimum temperature limits for the vegetative growth phase.

Faster development of non-perennial crops results in a shorter life cycle resulting in smaller plants, shorter reproductive duration and lower yield potential. Temperatures which would be considered extreme and fall below or above specific thresholds at critical times during development can significantly impact productivity. Photoperiod sensitive crops, e.g., soybean, would also interact with temperature causing a disruption in phenological development. In general, extreme high temperatures during the reproductive stage will affect pollen viability, fertilization, and grain or fruit formation. Chronic exposures to extreme temperatures during the pollination stage of initial grain or fruit set will reduce yield potential. However, acute exposure to extreme events may be most detrimental during the reproductive stages of development.

The impacts of climate change are most evident in crop productivity because this parameter represents the component of greatest concern to producers, as well as consumers. Changes in the length of the growth cycle are of little consequence as long as the crop yield remains relatively consistent. Yield responses to temperature vary among species based on the crop's cardinal temperature requirements. Warming temperatures associated with climate change will affect plant growth and development along with crop yield.

ii. Temperature extremes in climate

One of the more susceptible phenological stages to high temperatures is the pollination stage. Maize pollen viability decreases with exposure to temperatures above 35 °C. The effect of temperature is enhanced under high vapor pressure deficits because pollen viability (prior to silk reception) is a function of pollen moisture content which is strongly dependent on vapor pressure deficit. During the endosperm division phase, as temperatures increased to 35 °C from 30 °C the potential kernel growth rate was reduced along with final kernel size, even after the plants were returned to 30 °C. Exposure to temperatures above 30 °C damaged cell division and amyloplast replication in maize kernels which reduced the size of the grain sink and ultimately yield. Rice (*Orzya sativa* L.) shows a similar temperature response to maize because pollen viability and production declines as day time maximum temperature (T_{max}) exceeds 33 °C and ceases when T_{max} exceeds 40 °C. Current cultivars of rice flower near mid-day which makes T_{max} a good indicator of heat-stress on spikelet sterility. These exposure times occur quickly after anthesis and exposure to temperatures above 33 °C within 1–3 h after anthesis (dehiscence of the anther, shedding of pollen, germination of pollen grains on stigma, and elongation of pollen tubes) cause negative impacts on reproduction. Current observations in rice reveal that anthesis occurs between about 9 to 11 am in rice and exposure to high temperatures may already be occurring and will increase in the future. Given the negative impacts of high temperatures on pollen viability, recent observations suggest that flowering at cooler times of the day would be beneficial to rice grown in warm environments. They proposed that variation in flowering times during the day would be a valuable phenotypic marker for high-temperature tolerance. As daytime temperatures increased from 30 to 35 °C, seed set on male-sterile, female fertile soybean (*Glycine max* (L.) Merr.) plants decreased. This confirms earlier studiess on partially male-sterile soybean in which complete sterility was observed when the daytime temperatures exceeded 35 °C regardless of the night temperatures and concluded that daytime temperatures were the primary factor affecting pod set in soyabean. Crop sensitivity to temperature extremes depends upon the length of anthesis also. Maize, for example, has a highly compressed phase of anthesis for 3–5 days, while rice, sorghum (*Sorghum bicolor* L. Moench.) and other small grains may extend anthesis over a period of a week or more. In soybean, peanut (*Arachis hypogaea* L.) and cotton (*Gossypium hirsutum* L.) anthesis occurs over several weeks and avoid a single occurrence of an extreme event affecting all of the pollening flowers. For peanut (and potentially other legumes) the sensitivity to elevated temperature for a given flower, extends from 6 days prior to opening (pollen cell division and formation) up through the day of anthesis. Therefore, several days of elevated temperature may affect fertility of flowers in their formative 6-day phase or anthesis. Pollination processes in other cereals, maize and sorghum, may have a similar sensitivity to elevated daytime temperature as rice. Rice and sorghum have exhibited similar sensitivities of grain yield, seed harvest index, pollen viability and success in grain formation in which pollen viability and per cent fertility is first reduced at instantaneous hourly air temperature above 33 °C and reaches zero at 40 °C. Diurnal max/min day/night temperatures of 40/30 °C (35 °C mean) cause zero yield for those two species with the same expected response for maize.

The current evaluations of the impact of changing temperature have focused on the effect of average air temperature changes; however, increases in minimum air temperature may be more significant in their effect on growth and phenology. Minimum air temperatures are more likely to increase under climate change. While maximum temperatures are affected by local conditions, especially soil water content and evaporative heat loss as soil water evaporates, minimum air temperatures are affected by mesoscale changes in atmospheric water vapor content. Hence, in areas where changing climate is expected to cause increased rainfall or where irrigation is predominant, large increases of maximum temperatures are less likely to occur than in regions prone to drought. Minimum air temperatures affect night time plant respiration rates and can potentially reduce biomass accumulation and crop yield. Welch *et al.* (2010) found higher minimum temperatures reduced grain yield in rice, while higher maximum temperature raised yields; because the maximum temperature seldom reached the critical optimum temperature for rice. However, under the scenario of future temperatures increases, they found maximum temperatures could decrease yields if they are near the upper threshold limit.

Similar responses have been found in annual specialty crops in which temperature is the major environmental factor affecting production with specific stresses, such as periods of hot days, overall growing season climate, minimum and maximum daily temperatures, and timing of stress in relationship to developmental stages having the greatest effect. When plants are subjected to mild heat stress (1 °C to 4 °C above optimal growth temperature), there was moderately reduced yield. In these plants, there was an increased sensitivity heat stress 7 to 15 days before anthesis, co-incident with pollen development. Subjecting plants to a more intense heat stress (generally greater than 4 °C above optimum) resulted in severe yield loss extending to complete crop failure. Tomatoes under heat stress fail to produce viable pollen while their leaves remain active. The non-viable pollen does not pollinate flowers causing failure in fruit set. If the same stressed plants are cooled to normal temperatures for 10 days before pollination, and then returned to high heat, they are able to develop fruit. There are some heat tolerant tomatoes which perform better than others related to their ability to successful pollinate even under adverse conditions.

Perennial crops have a more complex relationship to temperature than annual crops. Many perennial crops have a chilling requirement in which plants must be exposed to a number of hours below some threshold temperature before flowering can occur. For example, chilling hours for apple (*Malus domestica* Borkh.) range from 400 to 2900 h (5–7 °C base) while cherry trees (*Prunus avium)* require 900 to 1500 h with the same base temperature. Grapes (*Vitis vinifera* L.) have a lower chilling threshold that other perennial plants with some varieties being as low at 90 h. Increasing winter temperatures may prevent chilling hours from being obtained and projections of warmer winters in California revealed that by mid-21st century, plants requiring more than 800 h may not be exposed to sufficient cooling except in very small areas of the central Valley. Climate change will impact the chilling requirements for fruits and nut trees also. Innovative adaptation strategies will be required to overcome this effect because of the long time requirements for genetic selection and fruit production once perennial crops are established.

Perennial plants are also susceptible to exposure to increasing temperatures similar to annual plants. These responses and the magnitude of the effects are dependent upon individual species. Exposure to high temperatures, >22 °C, for apples during reproduction increases the fruit size and soluble solids but decreases firmness as a quality parameter. In cherries, increasing the temperature 3 °C above the 15 °C optimum mean temperature decreases fruit set. Optimum temperature range in citrus (*Citrus sinensis* L. Osbeck) is 22–27 °C and temperatures greater than 30 °C increased fruit drop. During fruit development when the temperatures exceed the optimum range of 13–27 °C with temperatures over 33 °C there is a reduction in Brix (sugar content), acid content, and fruit size in citrus. Temperature stresses on annual and perennial crops have an impact on all phases of plant growth and development. The effects of temperature extremes on the plant could be from the combined effect of the warm air temperatures and the increasing atmospheric demand.

iii. Sum up

Temperature effects on plant growth and development is dependent upon plant species. Under an increasing climate change scenario there is a greater likelihood of air temperatures exceeding the optimum range for many species. Cool season species will have a constrained growing season because of the potential of average temperatures exceeding their range as illustrated by broccoli in Fig. 1. The effect of temperature extremes on plant growth and development has not been extensively studied with the major effect during the pollination phase. Exposure of plants to temperature extremes at the onset of the reproductive stage has a major impact on fruit or grain production across all species. One potential strategy to minimize this impact is to select varieties which shed their pollen in the early morning when temperatures are cooler.

The effects of increased temperature exhibit a larger impact on grain yield than on vegetative growth because of the increased minimum temperatures. These effects are evident in an increased rate of senescence which reduces the ability of the crop to efficiently fill the grain or fruit. Observations in controlled environment studies show that maize grain yield is greatly reduced by above normal temperatures during the grain-filling period. Temperature effects interact with the soil water status which would suggest that variation in precipitation coupled with warm temperatures would increase the negative effects on grain production. These observations and the previous results from the literature suggest that more research needs to be conducted to quantify the interactions between temperature and soil water availability across germplasm within a species and among species to determine potential adaptation strategies to offset negative effects of extreme temperature events.

D. Vertical distribution of temperature in earth's atmosphere

The temperature of the Earth's atmosphere is not identical across the Earth. It differs in spatial and temporal dimensions. The temperature of a place depends largely on the insolation received by that place. The interface of insolation with the atmosphere and the earth's surface creates heat which is measured in terms of 'temperature'. Temperature differs significantly at different heights relative to the Earth's surface and this variation in temperature characterizes the four layers that exist in the atmosphere. These atmospheric layers include: the Troposphere, Stratosphere, Mesosphere, and Thermosphere (Fig.26).

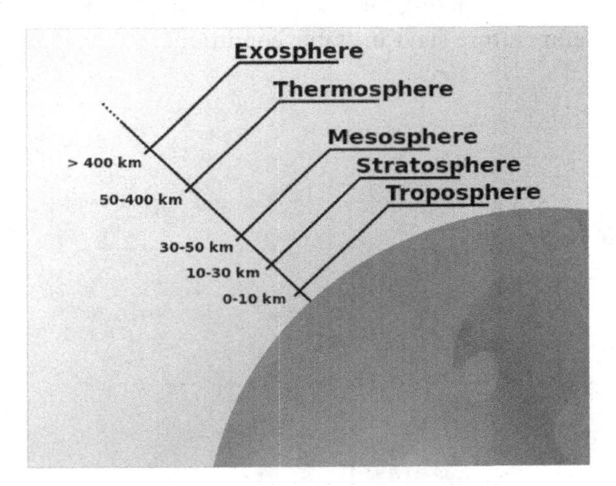

Fig.26: Vertical layers of atomosphere

a. **Troposphere:** The lowest most atmospheric layer, of the homosphere, is the thinnest of the layers, but it contains about 80 per cent of the mass of the atmosphere. The lowest most part of the troposphere is called the planetary boundary layer. Thickens of this layer is 7 to 8 kilometers in polar regions and as high as 16 to 18 kilometers in the tropics. There is also seasonal variation in the height. Where all type of climatic hazard or weathering phenomena are present e.g. fog, Clouds, due, frost, precipitation, thunderstorm, storms, cloud-thunder, lightning, Cyclones etc. In this layer Temperature decreases with increasing height at the rate of 6.45 °C per 1000 m. or 1 Km. This decrease of temperature rate is called "normal lapse rate". Top in the layer highest temperature during January and July over the equator, 45 °N and pole (North) is – 70 °C, 60 °C and 58 °C respectively. Occasionally the temperature is constant with height, or it may even increase with height in a thin layer called a "mechanical inversion" also called "negative lapse rater". When a dry air lifted upward due to subsidence of air, turbulence, convective and frictional forces mechanism (ascent or descent of air) without mixing heat with the environment, the lapse rate is approximately

10 °C per kilometer (5.5 °F per 1000 feet.), this cooling rate is called "dry adiabatic lapse rate". On the other hand, the latent heat of condensation released after precipitation is added to the ascending winds, with the result the temperature of the ascending winds decreases at the rate of 6 °C per Km. (3 °F per 1000 feet) this rate is called moist or retarded adiabatic rate.

Tropopause: The upper limit of the troposphere is called "Tropopause", literally means "Zone or region of mixing" or "where the mixing stops", which is about 1.5 km. thick. Here temperature stays in stable condition.

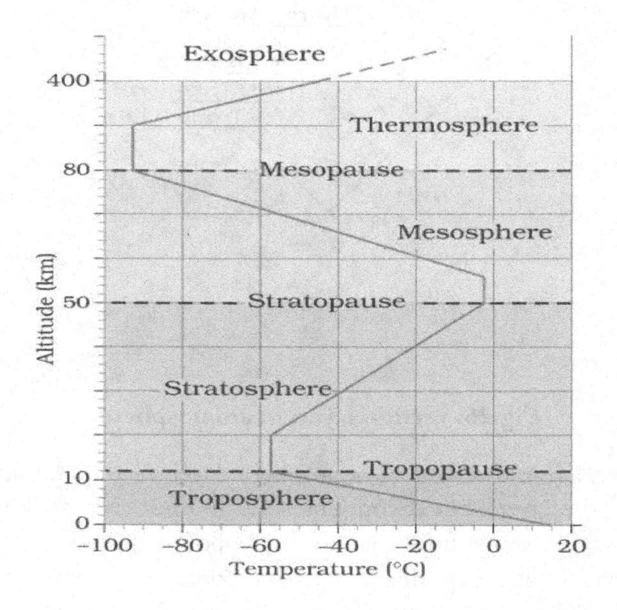

B. Stratosphere: This layer just situated above the tropopause, the average height of this layer is 25-30 km. so upper limit of the layer is 50 Km. There temperature is increase with increasing height, increasing rate is 5O C per Km., the temperature gradually rises upwards as it becomes 0 °C or 32 °F at the height of 50 Km. Lower part of the layer having maximum concentration of ozone (O_3) is called ozonosphere (15 – 35 Km.). The ozone layer absorbs most of the ultraviolet rays (A, B etc.) radiated from the sun and the thus the temperature of this layer becomes much higher than the other layers lying above and below ozone layer, it is called thermal upper air inversion.

Stratopause: Also its nature is similar to layer of tropopause. The nature of temperature of this layer is stable.

C. Mesosphere: This atmospheric layer extends between 50 km. to 80 Km. Temperature again decreases with increasing height. At the upper most limit of the layer temperature becomes – 80 ^{O}C to – 100 ^{O}C.

Mesopause: The upper most part of the mesosphere layer. There temperature nature is stable condition. After this thin layer temperature is continues rapidly increase with increasing height.

D. Thermosphere: The layer of the atmosphere beyond mesopause, wherein temperature increases rapidly with increasing height. The temperature at its upper limit (600 Km.) becomes 1700 ^{O}C. After this layer temperature is continues rapidly increase with increasing height. The temperature becomes 5568 ^{O}C at its outer limit (10,000 Km.) but this temperature is entirely different from the air temperature of the earth's surface as it is never felt.

Chapter - 15

Ecological Basis for Environmental Management

Historically, approaches to environmental management activities have been reactive rather than proactive. Environmental laws and regulations have been generated primarily in response to particular issues (*e.g.*, chemical contamination), creating a piecemeal approach for managing the environment. Responsibilities for managing different resources (*e.g.*, water, air, forests, wildlife) have been assigned to different agencies or groups within government, further fragmenting environmental management. Proactive approaches that recognize the interconnectedness of environmental components are necessary to address complex and long term environmental management issues. This perspective proposes an environmental management approach that is comprehensive and systematic, while still being comprehensible to decision makers and other stakeholders. The proposed approach is based on ecology and environmental values related to decision making. It considers interrelationships among and between living organisms (including humans) and their physical environment.

Humans have become the first species on Earth to interact as a whole with the global ecosystem. Humans influence and are influenced by ecosystems. Thus, humans are an integral part of today's ecosystems and fully depend on ecosystems for their well being. Yet as fundamental as this concept is, there is considerable tension regarding the role humans have in ecosystems and the latitude the human race should assume for manipulating ecosystems for its own purposes. Part of this tension comes from mixing short term and long term aspects of human/ecosystem interaction. In the short run, ecosystems provide goods and services, including livelihood, for many people. In the long run, ecosystems must persist if they are to provide the same opportunities ecosystem management is to balance the short term demands for products and services with the long term need for persistence.

Society has become concerned about the human condition and its relationship with ecosystems. We have learned that we cannot have a wise relationship with our environment by looking at it piecemeal or by ignoring the long term effects of our actions. As a species, we have out distanced our predators, drastically increased our numbers, and dominated many of our ecosystems, often using highly developed technology. Clearly, a number of global examples exist where ecosystems have been destroyed or severely damaged, leaving behind societies that struggle for subsistence. In the United States, the per capita rate of use of natural resources is one of the highest in the world. While natural resource utilization has benefited the economies of local communities, these benefits are offset by reductions in many important habitats and plant and animal species they support. This called the setup of number of professional and special interest groups and organizations, which are moving purposefully toward a more holistic form of managing ecosystems for long term sustainability.

The task is daunting and includes difficult issues such as air and water pollution and incompatible conterminous land use for which mitigation procedures are difficult at best. In many cases, humans not only depend on ecosystems, they also are the dominant stress to ecosystems. Much of the human stress to ecosystems stems from economic philosophy emphasizing a short term profit motive and from simple increases in population density, both of which impact resources and seriously challenge the concept of sustainability. When advanced technology is factored in, humans have exhibited great capacity to disrupt ecosystem processes.

While our study should focus on ecological characteristics rather than social and economic considerations, clearly the closer that ecological, social, and economic considerations are in agreement, the greater is the likelihood that both ecosystems and society will be sustainable (Fig. 1). Much past human impact lies outside the physical and biological capability of sustainable ecosystems. Much of this impact may have resulted from human wants for exceeding needs and the result has been a significant deterioration in many ecosystems.

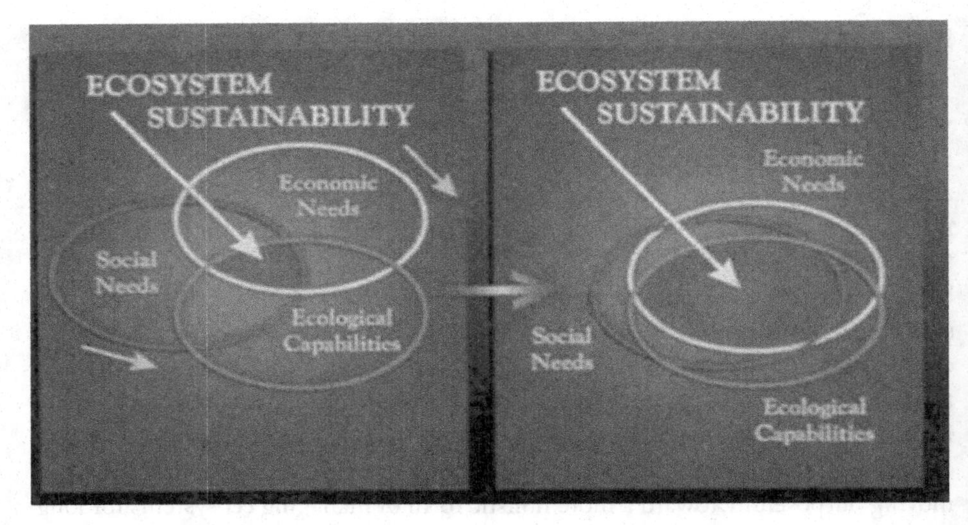

Fig.27: (Left) Relationship between ecosystem capabilities and social and economic needs; (Right) Relationship between ecosystem capabilities and social and economic needs with greater convergence, resulting in improved ecosystem sustainability. Because the physical and biological capabilities of ecosystems are limited in flexibility, convergence requires shifts in social and economic needs to comply more with ecosystem capabilities.

Ecosystem management is a logical step in the evolution of society's thinking and understanding about natural resource management. Ecosystem management involves a shift in focus from sustaining production of goods and services to sustaining the viability of ecological, social and economic systems now and into the future. This is brought about by bringing ecosystem capabilities and social and economic needs into closer alignment (Fig. 27 right). But ecosystems function sustainably only when they remain within normal bounds of their physical and biological environment. Thus ecosystem management will be successful only when management decisions reflect understanding and awareness of ecological principles related to sustainability. It is important to recognize that the human interest is served if long term ecosystem sustainability is assured, even if this requires altering certain human activities to stay within the physical and biological capabilities of ecosystems.

Ecosystem management may require new approaches such as less dependence on raw fiber, better utilization of existing natural resources and reduced human demands. Whether or not society has the capacity or fortitude to sustain ecosystems, efforts toward this goal probably will reduce the magnitude of long term social problems associated with ecosystem abuse. A way to approach ecosystem management is to identify underlying principles that apply uniformly regardless of the types of ecosystems being considered.

Guiding principles

In examining ecological aspects of ecosystem management, the Study Team formulated guiding principles based upon the fundamentals of conservation biology. These principles address portions of the ecosystem management principles outlined at a USDA Forest Service (1992a) workshop in Salt Lake City, Utah. The principles presented here, however, focus primarily on the ecological aspects of ecosystem management and are not intended to address all issues of managing ecosystems at the same level of detail. The guiding principles presented below are very similar to Grumbine's (1992) ecosystem management goals. While not all principles are universally accepted and may be found inadequate in certain cases, the Study Team concluded that the guiding principles outlined below address most of the biological problems associated with assuring ecosystem persistence for future generations, and hope that the principles will guide certain aspects of ecosystem management.

Principles of ecosystem management can be applied regardless of the degree of past or present human influence on the ecosystem. The basic idea is to view every action or change as an effect in a complex system of processes and to evaluate actions or changes from the perspective of the whole.

The assumption is that naturally evolving ecosystems (minimally influenced by humans) were diverse and resilient and that within the framework of competition, evolutionary pressures, and changing climates, these ecosystems were sustainable in a broad sense. Many present ecosystems modified by modern industrial civilizations do not have all these characteristics. The guiding premise for sustaining ecosystems and protecting biodiversity now and into the future is to manage ecosystems such that structure, composition and function of all elements, including their frequency, distribution and natural extinction are conserved. Conservation focuses on maintaining and restoring suitable amounts of representative habitats over the landscape and through time.

The following guiding principles expand our premise and provide an ecological basis for analysis and decision making:

1. Humans are an integral part of today's ecosystems and depend on natural ecosystems for survival and welfare; ecosystems must be sustained for the long term well being of humans and other forms of life;

2. In ecosystems, the potential exists for all biotic and abiotic elements to be present with sufficient redundancy at appropriate spatial and temporal scales across the landscape;
3. Across adequately large areas, ecosystem processes (such as disturbance, succession, evolution, natural extinction, re-colonization, fluxes of materials, and other stochastic, deterministic and chaotic events) that characterize the variability found in natural ecosystems should he present and functioning;
4. Human intervention should not impact ecosystem sustainability by destroying or significantly degrading components that affect ecosystem capabilities;
5. The cumulative effects of human influences, including the production of commodities and services, should maintain resilient ecosystems capable of returning to the natural range of variability if left alone; and
6. Management activities should conserve or restore natural ecosystem disturbance patterns.

Ecosystem management applies to all ecosystems, ranging from those having minimal human influence to those severely impacted by human activity. Ecosystem management should involve consideration of not only goods and services but also the viability of ecological, social, and economic systems now and in the future. Achieving this goal will require that ecosystem conditions, natural processes, natural disturbance patterns and productive capabilities be incorporated into decision making processes so that human needs are considered in relation to the sustainable capacity of the system. The principles reflect a need to embrace a land ethic that strives above all to sustain biological diversity and productive potentials of ecosystems. Furthermore, the principles may further encourage the distinction between human needs and human wants.

This approach would generally preserve all components of natural ecosystems, but it is not intended to revert all lands to a natural state. It does mean that management activities for ecosystems, regardless of the degree of human impact, must be within the physical and biological capabilities of the land, based upon an understanding of ecosystem function. It means saving all the components of ecosystems, including the structure, composition (including genetic diversity), and processes that characterize natural ecosystems. It means protecting and restoring the pieces of the landscape made uncommon by human activities, carefully reviewing existing impacts of nonnative species, and preventing the introduction of new ones.

Maintaining viable populations of all native animal and plant species is a central theme of ecosystem management, although major scientific knowledge gaps exist. For example, there is limited information identifying what minimum viable populations are. Ecosystem management also conserves soils, aquatic and riparian systems, and water resources. Ecosystem management cannot assure that rare animals and plants will reproduce and thrive, even though the protection of such species is a clear goal in many ecosystems. Ecosystem management is intended to allow normal fluctuations in populations that could have occurred naturally. It should promote biological diversity and provide for habitat complexity and functions necessary for diversity to prosper. It should not be a goal to maintain all present levels of animal populations or to maximize biodiversity.

Tools for maintaining viable species populations are likely to be focused on providing habitats in an appropriate spatial and temporal arrangement. Thus, vegetation management continues to be a major tool not only for commodity production, but also for maintaining and restoring biodiversity and for using habitat management to achieve delisting or to avoid listing of threatened and endangered species. However, management of other activities such as recreational use and management of exotic species may also be required.

Chapter - 16

Environmental Manipulation through Agronomic Measures

The environment of a crop constitutes both soil and aerial, as part of the plant is inside the soil and the other part is exposed to the aerial atmosphere. Environment influences growth and development of crops. Aerial environment includes solar radiation, rainfall, temperature, relative humidity and wind velocity. The aerial environment cannot be altered easily.

Soil environment is amenable for manipulation or modification (Fig.) through agronomic practices like tillage, irrigation, weeding, fertilizer application etc. The soil physical environment comprises of soil structure, soil air, soil water and soil temperature. Soil solution pH, electrical conductivity and nutrient concentration constitute the soil chemical environment. The soil biological environment constitutes the living organisms (micro-flora) present in the soil and the resultant reactions like mineralization, immobilization etc. The soil management part of agronomy deals with the manipulation of soil environment for better crop growth.

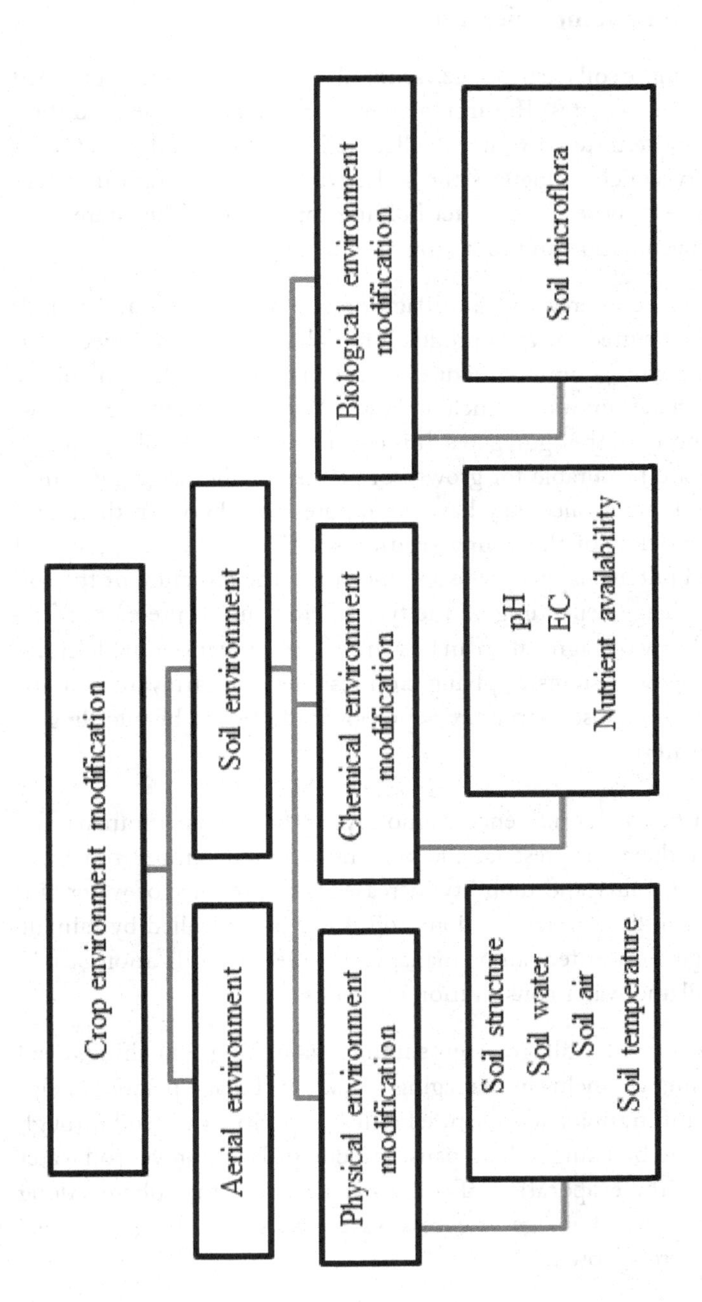

Fig.28: Crop environment manipulation

Soil physical environment modification

Soil structure: The primary soil particles viz., sand, silt and clay are usually grouped together in the form of aggregates. The arrangement of primary particles and their aggregates into certain definite patterns is called soil structure. Soil structure is an important property which influences the soil environment through its effect on the amount and size of pore space, water holding capacity etc. The altered soil environment affects germination and root growth of crops.

Soil structure can be altered by the tillage practices. Granular and crumb structure of soil are best suited for crop production. When the soil is subjected to tillage at optimum moisture, crumb structure is developed so that loss of soil by erosion is greatly reduced. Rain water is held in large pores, between the aggregates and also in the micropores of the aggregates. It is considered that all soil aggregates of 1 to 5 mm in size are favourable for growth of plants. Smaller aggregates may clog the soil pores and larger ones may have large pore space between them and the development of rootlets of the young seedlings. Soil structure is destroyed when tillage is carried out at inappropriate soil moisture. The structure of the soil is easily altered while tillage depending on the type of crop and its preference. The improvement and stability of aggregates can be achieved by growing suitable crops, adopting suitable cropping systems, applying crop residues and farmyard manure. But for paddy cultivation, the soil structure is purposely destroyed by puddling to avoid percolation of water.

Soil water: Soil water has major influence on modifying the soil environment. The soil strength is reduced due to its presence and moderates the soil temperature from extremes. Nutrient availability and mobility increases with presence of water also biological activities including mineralization. Soil water is controlled by rainfall, irrigation and drainage. Soil water can be managed by effective utilization of rain water by following soil and water conservation measures.

Agronomic practices like tillage, deep summer ploughing, mulching, dead furrows, contour farming etc. helps in managing soil water. Tillage practices helps a lot in increasing the infiltration and moisture holding capacity by the soil through increased pore spaces and breaking of hard pans. Mulching also improves soil water conditions by reducing the evaporation of soil water into the atmosphere. Along with retention of soil water, drainage of excess water also should be given equal importance for proper crop growth.

Soil air: The volume of the soil that is not occupied by soil particles is known as pore space. The pore space is usually occupied by air and water. Water and air compete for the same pore space and their volume fractions are so related that an increase of one generally decreases the other.

Soil air is virtually in continuation of atmospheric air and there is continuous exchange of gasses between atmosphere and soil air. Soil air differs from the atmospheric air by having a higher carbon dioxide, relative humidity and lower oxygen content comparatively. The circulation of air in the soil mass is known as soil aeration. The movement of soil air is influenced by temperature, water and diffusion. Soil air can be controlled mainly by providing good drainage facilities. The selection of suitable method of irrigation is necessary to avoid application of excess water. In case of submerged soils, most of the air escapes from the soil pores and air is present only in few pockets of subsoil.

Soil temperature: Soil temperature is one of the most important soil properties that affect crop growth. The major source of heat is sun and heat generated by the chemical and biological activity of the soil is negligible.

Control of soil temperature is difficult and expensive. However, some management practices slightly alter the temperature. Moisture content of the soil controls soil temperature to some extent. A soil that is over wet due to water logging and poor drainage has low temperature. In such soils, temperature can be raised by draining excess water. Water heats and cools slowly compared to soil. This principle is used to moderate soil temperature in lowland rice. About 10-30 cm of water is maintained in Japan during vegetative stage of rice to avoid adverse effect of low temperature. Plant canopy reduces soil temperature in warmer region. The seedlings of tea, coffee, tobacco etc., are raised under shade to avoid damage by high soil temperature. Mulching materials like straw of cereals reflect back some of the radiation and reduce soil temperature. Black materials like farmyard manure, charcoal, polythene film etc., absorb most of the radiation and increase soil temperature. Tillage influences soil temperature through its influence on aeration and bulk density. Tillage creates soil temperature optimum for seed germination and seedling establishment. Tillage loosens the soil surface resulting in decrease of thermal conductivity and heat capacity.

Soil chemical environment modification

pH: pH is defined as the negative logarithm of hydrogen ion activity. The pH of a soil indicates its acidity or alkalinity. pH influences nutrient availability, soil physical condition and plant growth.

» **Nutrient availability**: pH influences rate of nutrient release through its influence on decomposition, cation exchange capacity and solubility of materials. Decomposition of organic matter is slowest at pH below 6 and faster between 6 to 8. Availability of Fe, Al, Mn, Ca, Zn, P and B through its effect on solubility, leaching and precipitation with other nutrients.

» **Soil physical conditions**: Soil physical condition is related to pH and has an important effect on plant growth. Soil with pH above 8.5 is alkaline with high sodium content which deflocculates soil colloids. It results in destruction of soil structure and movement of water and air are impaired in the soil.

» **Plant growth**: High or low pH of nutrient medium has less influence on plant growth when provided with sufficient nutrients. However, soil pH influences plant growth through its effect on nutrient availability and soil physical condition.

Application of lime at full lime requirement increases the soil pH and the increase is gradual and seen upto 30 days after application of lime. Application of gypsum can reduce the soil pH. Apart from lime or gypsum application, agronomic practice of rice i.e., flooding can alter the soil pH. The pH of most acid and alkaline soils converges between 6 and 7 within 2 to 3 weeks after flooding.

Electrical Conductivity (EC): Electrical conductivity of soil is defined as the reciprocal of the electrical resistance of the extract of the soil which is one centimeter long and a cross-sectional area of one square centimeter. It is expressed as dSm at 25 °C and is used to express the salinity of the soil. Generally soils with cation exchange capacities have high electrical conductivity. EC can be mainly managed by leaching of excess salts and providing free drainage. Avoiding irrigation with salt water and less use of fertilizers which contribute to soil salinity also helps in decrease of soil electrical conductivity.

Nutrient availability: Increasing the rate of decomposition increases the availability of nutrients but, it is greatly influenced by the pH, EC, organic carbon content, microflora and edaphic factors. Soil microflora plays a key role in nutrient transformation. Agronomic practices like green manuring and crop rotation with pulses increases the availability of the soil nutrients. Flooding or submergence increases the availability of iron, manganese, sulfur, phosphorus and potassium and decreases the availability of zinc and copper. Under submerged conditions, most of the nitrogen will get lost through denitrification.

Soil biological environment modification

The soil biological environment is composed of several organisms belonging to both plant and animal kingdom. Because of their number and activity, several changes take place which are generally helpful to the plants while few of them are harmful. The activity and number of soil organisms can be modified by management practices so as to provide suitable environment for plant growth.

Soil microflora:

Soil microflora is mainly composed of bacteria, fungi and actinomycetes. Soil biological environment can be modified by cultural practices to provide suitable environment for crop growth. Tillage loosens the soils and incorporates plant residues thus increasing aeration and organic matter. As a result, microbial population proliferates and nutrients are released due to faster mineralization. Toxic chemicals like pesticides, herbicides and allelochemicals are also broken down faster due to better aeration. Irrigation provides sufficient soil moisture and increases the microbial population. Application of amendments bring soil pH to neutral. The favourable pH increases microbial population. Application of organic manures increases heterotrophic microflora. Oil cakes of neem, karanjia and groundnut increase the population of nematode trapping fungi. Neem cake is toxic to nematodes. Thus, these materials reduce nematode attack on the crop. Application of nitrogenous fertilizers decreases the population of bacteria and increases actinomycetes and fungi. Phosphatic fertilizer increases the population of nitrogen fixing bacteria.

The biological environment of submerged soils is quite different from that of aerated soils. The important organisms of submerged soils are Azolla, blue green algae, denitrifying bacteria and rhizosphere bacteria. Actinomycetes and fungi are absent in submerged soils. Non symbiotic bacteria present in anaerobic soils and those in the rhizosphere of rice roots fix atmospheric nitrogen. Nitrogen fixation in rice rhizosphere starts a month after transplantation and reaches peak at heading. These bacteria can fix N upto 63 kg/ha.

Chapter - 17

Improvement of Unproductive Lands Through Crop Selection and Management

Land is considered 'Unproductive' when it's productivity is diminished. Unproductive land means lands, including wetlands, which by its nature is incapable of producing agricultural or forest products due to poor soil or site characteristics, or the location of which renders it inaccessible or impractical to harvest agricultural or forest products.

From the total land area of 328 million hectare about 162 million hectare categorized as i.e. 51% is agricultural land, 4% is pasture land, 21% is forest land and 24% is wasteland.

Reasons for development of unproductive lands

» Land degradation

» Surface runoff and floods

» Shifting cultivation

» Soil erosion and desertification

» Loss of nutrients due to continuous erosion

» Soil acidification / alkalinisation

» Soil salinity

» Water logging

» Loss of biodiversity

» Long term socio-economic impact from humans like migration

Ways to improve unproductive lands

- » Planting and sowing of multi-purpose trees, shrubs, grasses, legumes and pastures.
- » Promotion of agro-forestry and horticulture.
- » Encouraging natural regeneration.
- » *In-situ* soil and moisture conservation measures like terracing, bunding, trenching, vegetative barriers and drainage line treatment.
- » Wood substitution and fuel wood conservation measures.
- » Awareness raising, training and extension education.
- » Encouraging people's participation through community organization and capacity building.
- » Drainage line treatment by vegetative and engineering structures
- » Development of small water harvesting Structures.
- » Afforestation of degraded forests and non forest wasteland.
- » Development and conservation of common property resources.

Different unproductive lands and their management

I. Salt-affected soils

Several factors contribute to the development of salinity and alkalinity like

- » Arid and semi-arid climate
- » High water table and impeded drainage
- » Impervious hard subsoil due to kankar pan
- » Basin shaped topography
- » Salt bearing substrata or use of brackish water for irrigation
- » Excessive canal irrigation or flooding

Reclamation techniques in salt affected soils

1. **Soil working:**

   a. Making the soil loose for rapid root development

   b. Breaking of kanker pan or clay pan

c. Leaching of soluble salts

d. Use of soil amendments

e. Use of phosphate fertilizer helps in mitigating adverse effects of salts on plants

2. Planting methods:

Successful afforestation of highly saline soils generally requires improvement in soil conditions through the application of appropriate planting techniques. Suitable methods of planting in saline soils are listed in table-1 and few other methods suitable for saline soils are also illustrated in fig.29.

Table 1: Planting methods on saline soils (Tomar, 1997)

Techniques	Size/height	Tools used	Amendments used
Pit method	45 x 45 x 60 cm	Spades	Gypsum, pyrite, manures
Sub-surface planting	15 cm (dia) and 45 cm (deep)	Auger	Soil and FYM
Ridge-trench method	40 cm (height)	Spade	FYM and fertilizers
Furrow planting	20 cm (deep) and 60 cm (wide)	Furrow maker	Soil and FYM

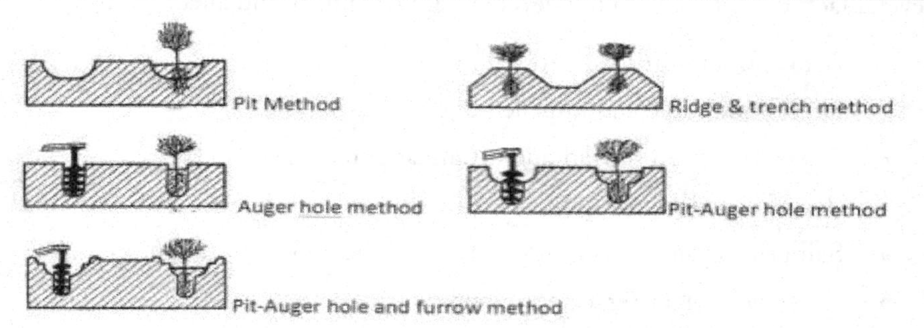

Fig.29: Methods of planting in saline soils

3. Silvi-pastoral model:

This model was developed at Karnal and comprises of growing *Prosopis juliflora* and karnal grass (*Leptochloa fusca*) together for about five years and then the grass is replaced by better forages such as berseem and senji. In about 8-10 years, due to the growth of trees and forages the soil will get reclaimed to an extent that a normal crop may be raised after removing tree or association with trees (Singh, 1995).

4. Choice of suitable species:

Normally firewood species are grown on salt affected soils. The species should be salt tolerant, having good soil binding characteristics with fast growing nature. However, in cultivable lands the following species can perform better.

- Cotton, sorghum, pearl millet *etc* should be grown during *kharif*
- Wheat, barley, mustard and sunflower during *Rabi* depending on the availability of irrigation water.
- Recommended cropping sequences for saline soils are pearl millet-barley, pearl millet-wheat, pearl millet-mustard and sorghum-wheat or barley.

II. Waterlogged wastelands

This problem occurs when soil pores within a few meters of the soil surface are saturated with water so that there is a rise of sub soil water. If the depth of water table is < 2 m then it is called as waterlogged area.

Water logging conditions can be categorized into:

- Water logging conditions caused by stagnant water
- Water logging conditions caused by impeded but mobile water

Management

1. **Drainage :** Draining of excess water from surface and subsurface layers
2. **Ridge and furrow method**: it is applicable where water depth is less than 1 meter. Height of ridge depends on the water depth. Furrow on the sides of the ridge help to drain excessive water. Planting is done on the ridges.
3. **Mound method:** In scattered patches of waterlogged areas and burrow pits along road, canal and railway lines, it is advisable to raise mounds about 1 meter at the base and 50 cm at the top. Tall plants are planted at the centre of the mound. It is necessary to use borrowed earth for mounds.

4. **Tall planting method:** In places where water stagnates for 2-3 months, tall plant can give fairly good success. Earth work is to done during dry October- November and planting frost hardy species so that by the next rainy season plants are established to combat the effect of water logging.

5. **Selection of suitable species:** Plants to be planted on these lands should have high transpiration rates and high osmotic potential (Table 2).

Table 2: Choice of suitable species for different degraded soils

Wastelands	Conditions	Suitable species	References
Saline soil	7-10 dSm^{-1} (EC)	*Acacia auriculiformis, A. deami, Salix wmini, S. fruticosum, Salix* spp, *Albinia guachepele, Tamarindus indica, A. lebbeck* and *A. caribea*	Tomar (1997)
	10-15 dSm^{-1} (EC)	*Pithecellobium dulce, Callistemon lanceolatus, A. nilotica, A. tortilis, Camarilla glauca, Eucalyptus tereticomis, A. catechu, Terminalia mjrma* and *Pongamia pinnata*	
	25-35 dSm^{-1} (EC)	*Tamarix articulate, T. troupii, Prosopis juliflora, Parkinsonia acrdeata* and *Acacia famesiana*	
Alkaline soil	7.5-9.0 pH (Low)	*Dalburgia sissoo, Morus alba, Grevillia robusta, Azedirachta indica, Tectona grandis, Populus deltoides, Emblica officinalis, Hardwickea binnata, Kajellea pinnata, Punica granatum* and *Aegle marmelos*	
	9.0-10 pH (Medium)	*Albizia lebbeck, Pongamia pinnata, Butea monosperma, Feronia limonia, Eucalyptus tereticornis, Sesbania sesban, Zizyphus mauritiana, Psidium guajava, Syzyium cumini, Achras japota* and *Cassia siamea*	
	> 10 pH (High)	*Prosopis juliflora, P. cineraria, A. nilotica, Casuarina equisetifolia, Tamarix articulata* and *Terminalia arjuna*	

Waterlogged areas	Waterlogged areas	*Salix* spp., *S. cuminii, Terminalia arjuna, Saccharum spontaneous, Vetiveria zizaniodies, Themenda* spp. and *Avistida* spp.	Luna (2006)
	Waterlogged saline soils	*Acacia farnesiana, Parkinsonia aculeata, Prosopis juliflora, Salvadora persica, S. oleoides* and *Tamarix* spp.	
	Marshy areas	*Barringtonia acutangula, Lagerstroemia speciosa, Pongamia pinnata* and *C. equisetifola*	
	Fresh water Swamps	*Diaspyros embryopteris, Pterospermum acerifolium, Bischofia javanica* and Canes	
	Muddy areas	*Avicinia officinalis* and *Manilkara littoralis*	
Ravine areas	Gully slopes	*Acacia catechu* and *Dalbergia sissoo*	Luna (2006)
	Gully beds	*Morus alba, Broussonetia papyrifera* and *Dendrocalamus spp.*	
	Saline and alkaline gully beds	*Acacia nilotica, Azadiractha indica, Albizia* spp. and *Prosopis juliflora*	
	Sandy beds	*Holoptelia integrifolia, D. sissoo, Dichostrachys nutan, D. strictus, P. juliflora, A. nilotica* and *A. indica*	
Coastal lands	Salinity areas	*P. chilensis, Pongamia glabra, Nipa fruticans, Salvadora persica* and *Juncus regidus*	Luna (2006)
	Wind and salt spray	*C. equisetifolia, A. auriculiformis, Atriplex numularia* and *Simaruba* spp.	
	Waterlogged areas	*P. juliflora* and *Avicennia officinalis*	
Hot Desert & Shifting Sand dunes	15-30 cm (Rainfall)	*P. juliflora, A. tortalis, A. senegal C. olligonum polygonodies, Zizyphus* spp. and *Lasiurus sindicus*	Luna (2006)
	30-40 cm (Rainfall)	*A. Tortalis, P. juliflora, P. chinensis, Tecomella undulate, Zizyphus* spp., *Ricinus cumunis, Crotolaria burhia, Aerva javanica, Lasiurus subuducya, Cenchrus setigerus* and *Saccharum bengalensis*	
	> 40 cm (Rainfall)	*A. nilotica, T. undulate, parkensonia aculata, A. excalsa, A. lebbeck, R. cumunis, Ziziphus* spp, *Cassia auriculata, C. ciliaris, Panicum antidotale* and *S. bangalensis*	

Mined Spoil Areas	Bauxite mined areas	*Grevillea pteridifolia, E. camaldulensis, Pinus* spp. and *Shorea robusta*	Solanki (1999)
	Coal mines areas	*E. hybrid, E. camaldulensis, A. auriculiformis, A. nilotica, D. sissoo* and *Pongamia pinnata*	
	Lime stone mine spoils	*S. tetrasperma, Leucaena leucophala, Bauhinia retusa, A. catechu, Ipomia cornia, Eulalopsis binata, Agave americana, Pennnisetum purpureum* and *Erythrina suberosa*	
	Rock phosphate mine	*Saccharum spontaneum, Vitex negundo, Rumex hustatus, Mimosa himalayana, Buddlea asiatica* and *L. lucocephala*	
	Lignite mined area	*Eucaluptus* spp., *L. leucocephala, Acacia* spp. and *Agave* spp.	
Denuded Hill Slopes	Direct sowing method	*Pinus roxburghii, P. wallichiana, Cedrus deodara, A. dealbata, A. altissima, Cupressus torulosa* and *C. arizonica*	Luna (2006)
	Transplants/ Potted plants method	*C. deodara, C. torulosa, Morus* spp., *Pinus* spp., *Juglans regia* and *Robinia pseudoacacia*	
	Branch cutting method	*Acer negundo, Salix* spp., *Plantanus orientalis* and *Populus* spp.	
	Root suckers method	Ailanthus altissima	
Landslips & Landslides	Grasses	*Pennisetum* spp, *Arundo donex* and *S. munha*	Luna (2006)
	Shrubs	*Ipomea carnea, Vitex negundo, S. tetrasperma, Indigofera heterarantha, Cotoneaster, microphylla, Lantana camara, Dodonea viscose* and *Coriar nepalensis*	
	Trees	*P. longifola, Juglans regia, P. gamblei, Salix* spp. *Robinia pseudoacacia* and *Erythrina suberosa*	

III. Ravine lands

The word 'ravine' means a deep valley or gorge and associated with a system of gullies, running more or less parallel to each other. They represent the large stage of water erosion. Ravine area is characterized by stiff clay with poor water retention power, alkaline pH, more soluble salt, less soil fertility.

Developmental techniques

1. **Mechanical structures**: Such as terraces and gully control structures in general and often dams are constructed in catchments areas where runoff volume is very high.

2. **Sod-strip checks:** On gentle sloping beds these are used for arresting silts. The strips should have a minimum width of 30 cm and an escapement of 2 to 3 m from one another.

3. **Contour trenching:** On moderate slopes, contour bunding or trenching are done. Excess water is collected in catch water drains and diverted into natural water channels. Such drains should be dug at a minimum distance of 3 times the depth of the gully from its head. Easing of the head of the gully is done by cutting and filling in order to prevent water from scouring toe of the gully. The slope fall is tufted with grasses like *Cynodon*, *Amphilophis*, etc.

4. **Terrace or check dam:** Steep gullies are stabilized with building of check dam across them. Siltation fills in gullies and makes them into useable land again. In the deposited silt, grasses like *Pennisetum* and *Panicum* should be planted.

5. **Choice of suitable species:** The species should be drought and frost resistance (Table 2).

6. **Aerial seeding:**

» It is used in the state of Maharastra, Rajasthan, UP and MP for ravine afforestation.

» Seeds of *P. chilensis A. nilotica*, *Azadiractha indica*, *Albizia* spp, *D. sissoo*, *D. strictus*, etc. at the rate of 20-40 kg/ha were sown using Beaver and Dakota aircraft.

Success story of bamboo and anjan grass based agroforestry system for enhancing the productivity of ravines

In Gujarat, where dairy farming is one of the major enterprises, the demand for fodder is increasing day by day. This has put pressure on marginal lands and warrants putting these waste lands into productive use for sustained and regular supply of fodder. These non-arable lands in general, and ravine wastelands in particular, have potential to supply the much needed fuel and fodder. Therefore, degraded ravine land must be put to productive use to obtain economic returns while protecting the environment.

An indigenous technology for enhancing productivity of ravines has been evolved at Dehradun based Central Soil and Water Conservation Research and Training Institute (CSWCR&TI). Research centre located at Vasad in Anand district, Gujarat, which has been adopted by various developmental organizations. The centre has successfully developed a Bamboo and Anjan Grass based silvi-pasture system for enhancing productivity of ravines. Ravine lands, under unproductive use can be successfully reclaimed by planting Bamboo, a fast growing plant species, on the gully beds and Anjan grass on the side slopes and the interspaces of gully bed for economic utilization of gullied land. This technology gives high returns and also checks water erosion, thereby preventing soil loss from ravines. Further, reclamation of ravine lands not only provides livelihood support but also helps natural resource conservation and carbon sequestration in the long run. This technology was used for reclamation of nearly 1000 hectare of community and waste lands in Mahi river stretch in Gujarat.

The cost of planting bamboo and sowing grass slips in ravine works out to be about Rs. 22000/- per hectare. During the initial years, green fodder of anjan grass can be harvested from the degraded ravines every year. Planting of grass also protects the ravine slopes and reduces soil loss and about 7.1 t/ha/year of green fodder can be obtained from the stabilized slopes. The grass yield from inter spaces of bamboo planted on bed portion is about 10 t/ha/year during initial 5 years. The grasses fetched an income of Rs. 3000 to 6000/- per ha over the period of 5 years at Vasad. About 300 clumps of bamboo with 3000-4000 old and 1000 to 1500 new culms are available in one hectare bamboo plantation of over seven years old. About 30 per cent of old culms can be harvested easily per year. The bamboo planted on the ravine bed can fetch Rs. 6000 to Rs. 27000/ha/year (2008 prices). The watershed with this silvipasture system absorbs more than 80% of rainfall that is either utilized by the plant or percolated deep, to recharge the ground water. Due to less runoff, soil loss is reduced to less than one tonne per hectare per year only from about 20 tonne per hectare per year from degraded ravines prior to plantation.

A ravine is a landform narrower than a canyon and is often the product of stream cutting erosion. About 3.67 million hectare constituting 1.12% of total geographical area in the country is under ravine lands. In Gujarat, ravines occupy an area of 0.4 million hectare and extends from the south bank of the Tapti, Narmada, Watrak, Sabarmati and Mahi river basins. The productive utilization of these ravine lands for fodder production and commercial cultivation of bamboo will not only meet the fodder needs but also provide a supplementary source of regular income to the farmers.

IV. Coastal lands

Coastal lands are found in India in the form of narrow strips along the eastern and western coast. Large quantity of sand gets accumulated as a result of tide and high prevalent winds. The soil in coastal areas is purely sand which is unstable and has low water holding capacity, alkaline nature, poor nutrient content, high water table and poor textured soil.

1. **Fore-dune system:** Before plantation on the coastal sand dunes artificial dunes should be erected along the sea face to act as a barrier to prevailing strong winds. Artificial dune of 10-13 m high would generally be sufficient. Wooden planks like rejected railway sleepers or used telephone poles can be used for creating foredunes. After that *Ammophila arenaria* could be planted on foredunes.

2. **Live fences:** This technique is to trap the blowing sand with the use of sand fences. Sand fences slower the sand movement by reducing the wind velocity in their immediate vicinity. While sand fences are very effective in trapping windblown sand, once they are filled they have little or no further effect on sand movement. In India, live hedge of *Agave* and *Aloe* on boundary and *Ipomoea biloba* as a sand cover plantation have been successful.

3. **Dune grass mats and netting:** These are also useful techniques for protecting bare sand surfaces. Coarse netting and mats are useful in protecting dunes while transplanted dune grasses are establishing. This technique does protect the sand surface but does not collect much sand, so the best use of netting and mats is to protect new seedlings. Grasses are *Ammophila breviligulata, A. arenaria, Panicum amarum, Uniola paniculata* and *Spartina patens.*

4. **Lupin sowing:** Lupin (perennial legume herb) not only helps develop better protective vegetation but also adds nitrogen to the soil. Planted dune grass and other areas where lupin is absent receive a broadcast of 10-11 kg/ha of lupin seed at the time of the second fertilizer application in autumn.

5. **Tree planting:** Tree planting can begin after some 3 to 5 years of lupin growth. These plants must be able to tolerate rapid sand accumulation, flooding salt spray, sandblast, wind and water erosion, wide temperature fluctuations, drought, and low nutrient levels. Ridge method, sunken method

and raised mounds method are used in saline areas, wind and salt spray areas and waterlogged areas respectively (Table 2).

V. Hot desert and shifting sand dunes

The desert is dry, barren, hot, and silent with no a drop of water for months. Deserts are classified by their geographical location and dominant weather pattern. Desert soil is characterized by very low organic matter, high soluble salts, low nutrients status, high pH, poor soil structure and low water holding capacity.

Types of dunes:

» Crescent dunes formed by one dominant wind direction

» Linear dunes by the action of two wind direction

» Pyramidal dunes occurs in arid region without a dominant wind direction

Improvement techniques

1. **Mulching:** Soil water and oil by products (Bitumen emulsion) have been used as mulch for stabilizing moving dunes. Crude oil has been used in Arabian countries. It is heated to 50 °C and spread at the rate of 4 m^3 per hectare. It gives protection for 3-4 years which is enough for growth of vegetation.

2. **Raising of micro-wind breaks:** It is based on the principle of reducing the threshold wind velocity at the dune surface. Twigs of brushwood species such as *Calligonum poligonodies, Leptadinia pyrotechnica, Zizyphus nummularia, Aeurva persica, Crotolaria burhea* and *Panicum spp.* are inserted in the ground and planted in parallel row of 5 m spacing. This should be done before the monsoon season.

3. **Planting of cuttings:** Cuttings of 60-80 cm length is set into sand with aid of an iron bar, at least 50-70 deep. Deep planting protect the planting stock against deflection and secures root development in a comparatively moist sand layer. e.g. a. Pre sprouted branchs cuttings of *Tamarix articulata* and *Calligonum polygonoides.* b. Pre-sprouted stumps of *P. juliflora, A. lebbeck* and *D. sissoo.*

4. **Seeds spray method:** Seeds of grasses and legume crops mixed with clay and sodium arsenate are sown on the micro wind breaks side during monsoon season.

5. **Using of earthen bricks:** Planting of seedlings raised in earthen bricks in pits of 45 cm × 60 cm has proved successful re-vegetation on arid regions.

6. **Choice of suitable species:** Species for dune fixation must be highly drought resistant with well developed root system and should be wind firm.

Afforestation effects on soil quality of sand dunes

- » Improvement in physical, chemical and mineralogical properties of soil
- » Organic matter accumulation in the soil profile
- » Contents of total, organic and available nutrients in surface horizons are higher
- » Fixation of colloidal materials in soils such as humic acid and fulvic acid

VI. Cold deserts

Large area of Leh and Kargil in Jammu and Kashmir and some area of Lahul and Spiti valley in Himachal Pradesh fall in this category. Dryness is of two types here, one is caused by low precipitation inhibiting absorption of water by plant roots and the other caused by the dryness of the atmosphere.

Characteristics of cold desert

- » Total precipitation does not exceed 400 mm.
- » Soil vary from sand and sandy loam to loams
- » Topography is flat to undulating and stony rocks, river banks etc.
- » Neutral slight alkaline in reaction
- » Short growing period
- » Wind erosion is there

1. **Soil working:** Soil working is done during March-April followed by planting at April end. Soil working will depend upon type of land (Table 3).

2. **Vegetation:** Choice of species is limited to few genera *i.e., Salix, Populus, Juniperus,* Planting is done by stem cuttings. Standard cutting size is 0.25 cm to 7.5 cm in girth and 2 m to 2.5 m in length. In between the rows Alfalfa is

sown for fodder and it improve the soil texture. Planting is done at the spacing of 2 × 2m or 3 × 3 m.

Table 3: Soil working method for cold desert areas (Luna, 2006)

Methods	Conditions	Remarks
Trench-cum-pit type	Gently sloping areas with stony surface	Digging staggered contour trench spaced 10×10 m. In interspaces 60×60×100 cm pits spaced at 2×2 m are dug and both are filled with mix of soil + fine textured silt
Irrigation-cum-drainage type	Low lying areas with a high salt concentration	Planting is done in 45-50 cm deep holes. Make 45 cm high mound around the plants.

VII. Mined spoils

In India total mining wastelands is about 505.35 km^2 which is 0.02 % of total geographical area spread over 19 states. Maximum number of mines being in Jharkhand, MP, Rajasthan, Bihar, Orissa etc. The total area under mining in the country is equivalent to 1/3rd of that under agriculture. Establishing of vegetation on wastelands is slight difficult due altered pH, variable texture, lake of organic matter, fragmented rock and many other toxic chemical factors.

1. **Soil working:** Following techniques may be used to afforest mined areas

    b. Covering toxic or infertile material with soil or waste of better quality

    c. Neutralizing strong acid alkaline soil by the use of lime, sulphur or waste of opposite reaction

    d. Buffering toxic elements by the use of peat, humus, clay with a high exchange capacity

    e. Leaching the salts by the means of rain water

    f. Fertilizing with green manuring. Stabilizing the surface of wastelands by agglutination sprays or mulches

2. **Natural colonizers** such as *Deschapsia caespitosa, Poa alpinaperform* and *Phleum*

alpinum better than introduced species and respond to fertilizers. *Cynodones* spp, *Typha* spp and *Cyperus* spp are plants to naturally colonize the edges of drainage of tailing dams. Some natural colonizers are given in table 4.

Table 4: Natural colonizers for different mined areas

Species	Mined Areas	Remarks
Rumex acetosella	Al and Mg mines areas	P deficient soil
Atriplex nummularia	Ni mines	Alkaline soil
Atriplex canescens	Uranium and coal mines	
Robinia pseudoacacia	Lime, Coal, Fe, Sulphure	Acidic Soil
Quercus rubra	Brick wastes	
Populus tremula, P. nigra	Sand wastes	Drained soil
Salix spp.	Fe, Cu mined areas	
Betula spp.	Rock-phosphate, asbestos wastes	Acidic Soil

3. **Use of bio-fertilizers:** Use of beneficial microorganisms like VAM, *Rhizobium*, Azotobacter, phosphate solubulizing microorganisms and blue green algae etc are useful in the reclamation programme of mining areas.

4. **Agro-engineering measures:** It refers to construction of a series of gabian structure and loose rock filled check dams. In between trees, grasses and legumes can be successfully introduced and maximum benefits can be obtained from such areas.

5. **Use of micro-organisms and other living organisms:** Use of algae, lichens and mosses in establishment of tailing and other overburdens accelerates the establishment of higher species. Mycorrhizae which are not sensitive to heavy metals are one of the important techniques. Introducing earthworms, termites and other soil living insects into the cycle of reclamation to improve the rate of organic matter and porosity in soil has been suggested as a possibility in building up the top soil.

6. **Selection of species:** For re-vegetation of mine spoils, plant species should have N fixing ability, hardiness, quick coverage during early stage, ability of phyto-extraction and phyto-stabilization. Establishment of permanent vegetation cover of suitable trees and grasses mixtures will improve their deteriorated conditions and stop further degradation (Table 2).

7. **Hydro-seeding:** Sowing of seeds directly or by transplantation on difficult surfaces through the use of water called hydro-seeding. Large volume of water, seed, fertilizer and mulch are needed to apply in a fine spray. Any shredded organic material, such as bark, sawdust or hay can be used to establish proper microclimate and to avoid clogging and soil crusting. This technique is best suited particularly for grasses.

VIII. Denuded hill slopes

Generally found in subtropical and temperate region of country. The southern aspect of Himalaya is generally denuded and more common in western Himalaya than eastern. Because eastern part in more moist. Cause of denudation is unrestricted felling, excessive grazing and frequent fires, *etc.* because of which the soil is poor in nutrients and moisture.

1. **Soil working:** Soil working should be necessary before plantation of trees on any denuded hills slope. Two methods viz., contour trenching and pit method/thalis method are used. Contour trench with the size of 50×50 cm in cross sections is made on light sloppy areas. Whereas, a pit of 30×30×30 cm or 45×45×45 cm are made on steeper sloppy areas in pit method/thalis method.

2. **Planting methods:** Species which are propagated by vegetative parts are likely to succeed. Species should be frost hardy. In dry area of north-western India such species as *A. modesta, A. catechu, P. chilensis etc.* and in the peninsular region, *E. tereticornis, Anacardium occidentale, A. auriculiformis, C. siamia, A. lebbeck* etc. have been successfully used.

IX. Land slips and land slides

Downward or outward movement of slope forming material soil, natural rocks, vegetation etc. Land slides are vast problem in Northern and Eastern mountainous regions of India.

The main causes of landslides are

» Construction of roads,

» Over-grazing,

» Deforestation,

» Blasting,

» Earthquakes and

» Extension of agricultural activates in sloppy hilly areas

An avalanche, similar in mechanism to a landslide, involves a large amount of ice, snow and rock falling quickly down the side of a mountain.

Developmental techniques

1. **Wattling of area**: Wall lining is done by providing contour tranches 30 cm wide and 1 m deep which are filled with wood bundles. On the downhill slope posts of species which spread on planting such as *Lanna grandis, Erythrina subrosa, vitex nugundo, salix tetrasperma, Hamiltonia suaveolens* are placed in the trenches

2. **Mulching:** A thick layer (5-7 cm) of locally growing grass *Chrysopogon* spp., *Heteropogon* spp. or others are spread all over the outer side of land slips and binds with wire to prevent wind or runoff water displacing the mulch out of position.

3. **Planting of trees:** In order to permanently stabilize the slip area afforestation in essential and it involves two factor:

   a) Immediate covering the soil with quick establishing species

   b) Permanent cover to restore the original habitat conditions

Species should be vegetatively propagated with massive tap root system and climbing nature. They can be planted during winter or onset of monsoon. A hole is made in the soil or rock surface with a planting rod and cutting of about 30 cm length inserted and compacted.

Conclusion

It is realized that the balance between the land degradation and restoration rates should be maintained so as not to further degrade the land. Wasteland afforestation is found to be a financially viable and environmentally sound use of most of those lands. Unproductive lands can be suitably reclaimed for agriculture or some alternate uses following afforestation, agroforestry, and bio-engineering measures which are simple and cost-effective.

Farmers should be helped by the government officials for leasing of the land, credit, agricultural technological inputs, market information and marketing, including all the related materials regarding the wastelands reclamation. Just like the green revolution there should be programme for greening the Unproductive land and a coordinating agency at the state/district/block and village level should be established to help to improve unproductive land.

Chapter - 18

Effect of Climate Change on Crop Production

Climate change and variability are concerns of human being. The recurrent droughts and floods threaten seriously the livelihood of billions of people who depend on land for most of their needs. The global economy is adversely being influenced very frequently due to extreme events such as droughts and floods, cold and heat waves, forest fires, landslips etc. The natural calamities like earthquakes, tsunamis and volcanic eruptions, though not related to weather disasters, may change chemical composition of the atmosphere. It will, in turn, lead to weather related disasters. Increase in aerosols (atmospheric pollutants) due to emission of greenhouse gases such as Carbon Dioxide due to burning of fossil fuels, chlorofluorocarbons (CFCs), hydrochlorofluorocarbons (HCFCs), hydrofluorocarbons (HFCs), perfluorocarbons (PFCs) etc., Ozone depletion and UV-B filtered radiation, eruption of volcanoes, the "human hand" in deforestation in the form of forest fires and loss of wet lands are causal factors for weather extremes. The loss of forest cover, which normally intercepts rainfall and allows it to be absorbed by the soil, causes precipitation to reach across the land eroding top soil and causes floods and droughts. Paradoxically, lack of trees also exacerbates drought in dry years by making the soil dry more quickly. Among the greenhouse gases, CO_2 is the predominant gas leading to global warming as it traps long wave radiation and emits it back to the earth surface. The global warming is nothing but heating of surface atmosphere due to emission of greenhouse gases, thereby increasing global atmospheric temperature over a long period of time. Such changes in surface air temperature and consequent adverse impact on rainfall over a long period of time are known as climate change. If these parameters show year-to-year variations or cyclic trends, it is known as climate variability.

However, the official definition by the United Nations Framework Convention on Climate Change (UNFCCC) is that climate change is the change that can be attributed directly or indirectly to human activity that alters the composition of the global atmosphere and which is in addition to natural climate variability observed over comparable time periods. However, scientists often use the term for any change in the climate, whether arising naturally or from human causes. In particular, the Intergovernmental Panel on Climate Change (IPCC) defines climate change as a change in the state of the climate that can be identified by changes in the mean and 2 / or the variability of its properties, and that persists for an extended period, typically decades or longer.

Weather and climate: Weather is the set of meteorological conditions such as wind, rain, snow, sunshine, temperature, etc. at a particular time and place. By contrast, the term climate describes the overall long-term characteristics of the weather experienced at a place. The ecosystems, agriculture, livelihoods and settlements of a region are very dependent on its climate. The climate, therefore, can be thought of as a long-term summary of weather conditions, taking account of the average conditions as well as the variability of these conditions. The fluctuations that occur from year to year, and the statistics of extreme conditions such as severe storms or unusually hot seasons are part of the climatic variability.

The Earth's climate has varied considerably in the past, as shown by the geological evidence of ice ages and sea level changes, and by the records of human history over many hundreds of years. The causes of past changes are not always clear but are generally known to be related to changes in ocean currents, solar activity, volcanic eruptions and other natural factors. The difference now is that global temperatures have risen unusually rapidly over the last few decades. There is strong evidence of increase in average global air and ocean temperatures, widespread melting of snow and ice, and rising of average global sea levels. The IPCC Fourth Assessment Report concludes that the global warming is unequivocal. Atmosphere and ocean temperatures are higher than they have been at any other time during at least the past five centuries, and probably for more than a millennium. Scientists have long known that the atmosphere's greenhouse gases act as a blanket, which traps incoming solar energy and keeps the Earth's surface warmer than it otherwise would be, and that an increase in atmospheric greenhouse gases would lead to additional warming.

Important weather extremes and their impact at global level

The year 1998 was the warmest and declared as the weather-related disaster year. It caused hurricane havoc in Central America and floods in China, India and

Bangladesh. Canada and New England suffered heavily due to ice storm in January while Turkey, Argentina and Paraguay suffered with floods in June 1998. In contrast, 3 huge crop losses were noticed in Maharashtra (India) due to un-seasonal and poor distribution of rainfall during 1997-98. The 1997/1998 El Nino event (The El Nino is nothing but warming of Pacific), the strongest of the last century, affected 110 million people and costed the global economy nearly US$ 100 billion.

The year 2003 was the year of heat and cold waves across the world. The European Union (EU) suffered to a large extent due to heat wave that occurred in summer 2003. In India Uttar Pradesh, Bihar, West Bengal, Orissa and Andhra Pradesh are the States that experienced summer heat waves. When the EU suffered heat wave during the summer in 2003, India experienced severe cold wave from December 2002 to January 2003. Some parts of Jammu, Punjab, Haryana, Himachal Pradesh, Bihar, Uttar Pradesh and the North Eastern States experienced unprecedented cold wave. The crop yield loss varied between 10 and 100% in the case of horticultural crops and seasonal crops. The fruit size and quality were also adversely affected in horticultural crops. However, temperate fruits like apple, perch, plum and cherry gave higher yield due to extreme chilling. The damage was more in low-lying areas where cold air settled and remained for a longer time on the ground (Samra *et al.*, 2004).

High temperature in March 2004 adversely affected crops like wheat, apple, mustard, rapeseed, linseed, potato, vegetables, pea and tea across the State of Himachal Pradesh in India. The yield loss was estimated between 20% and 60% depending upon the crop. Wheat and potato harvest was advanced by 15-20 days and the flowering of apple was early by 15 days. The optimum temperature for fruit blossom and fruit set is 240 C in the case of apple while it experienced above 26 ^{0}C for 17 days. The entire region recorded between 2.1 and 7.9 ^{0}C higher maximum temperature against the normal across the State of Himachal Pradesh in March 2004 (Prasad and Rana, 2006). A decline of 39% in annual cocoa yield was noticed in 2004 when compared to that of 2003 due to rise in maximum temperature of the order of 1 to 3 ^{0}C from 14th January to 16th March in Central part of Kerala, India. Such trend was noticed whenever summer temperature shot up by 2 to 3 ^{0}C when compared to that of normal maximum temperature of 33 to 36.5 ^{0}C.

Untimely rains and hailstorms destroyed wheat crop of 15,000 hectares over UP, Haryana and Punjab in Rabi season 2007 in India. In contrast, heavy 4 snowfall over Kashmir valley was recorded in 2007 due to western disturbances. Similar was the case during monsoon 2007, causing floods across several continents (Hurricane Dean in August in Mexico) including India and Bangladesh. Heavy rains again in

September in Andhra Pradesh, Karnataka and Kerala led to floods and thus the year 2007 was declared as the flood year in India. A huge crop loss was noticed in several states of the Country due to floods in *kharif*, 2007. Similar was the case in Algeria, Uganda, Sudan, Ethiopia and Kenya. Bangladesh suffered heavily due to super cyclone ' Sidr' that hit in November 2007.

Beijing in China had temperature as high as 16 ^{0}C in February 2007, the highest since the meteorological record began in 1840, followed by one of its coldest and snowiest winter in 2008. As a result of heavy snow for a period of three weeks since 10th of January 2008, 104 million ha. of farm land was damaged in addition to destruction of 22,000 houses and the economic loss was estimated at $ 7.5 billion. The La Nina phenomena may be another reason for severe snow storms. The La Nina is a large pool of unusually cold water in the equatorial Pacific that develops over a few years and influences global weather, which is opposite to El Nino. The El Nino is nothing but warming of Pacific. The mercury dipped to a new low of 9.4 ^{0}C over Mumbai on 6.2.2008. The frequency of such unusual weather phenomena is likely to increase across the world and huge economic loss is expected.

The Mean Sea Level (MSL) rise is likely to be slightly less than one mm/year along the Indian coast. Sea level rise may lead to disappearance of low-lying areas of coastal belt in addition to changes in ocean biodiversity. It threatens health of corals and polar bear population. Greater number of high surges and increased occurrences of cyclones in post-monsoon period, along with increased maximum wind speed, are also expected. This phenomenon of sea level rise threatens the area of land available for farming.

As per the United Nations Report of FAO, India stands to lose 125 million tonnes equivalent to 18% of its rainfed cereal production from climate change by 2015. China's rainfed cereal production potential of 360 million tonnes is expected to increase by 15% during the same period. It would also cause a worldwide drop in cereal crops, leaving 400 million more people at risk of hunger, and leaving three billion people at risk of flooding and without access to fresh water supplies. The crop 5 production losses due to climate change may also drastically increase the number of undernourished people, severely hindering progress in combating poverty and food security. The severest impact is likely to be in sub-Saharan African countries, which are least able to adapt to climate change or to compensate for it through increase in food imports. In 2004 and 2005, twenty four (24) sub- Saharan African countries faced food emergencies, caused by a lethal combination of locusts and drought. In addition, adverse hot and dry weather in United States and drought conditions in parts of the EU lowered cereal output during 2005 when compared to that of 2004.

The simulation models indicate that the global warming leads to reduction in rice and wheat production in northern India.

The Indian economy is mostly agrarian based and depends on onset of monsoon and its further behaviour. The year 2002 was a classical example to show how Indian food grains' production depends on rainfall of July and it was declared as the all-India drought, as the rainfall deficiency was 19% against the long period average of the country and 29% of the area was affected due to drought. The "All India drought" is declared when the rainfall deficiency for the Country as a whole is more than 10% of normal, and when more than 20% of the Country's area is affected by drought conditions. The *kharif* season food grain production was adversely affected by a whopping fall of 19.1% due to "All–India drought" during monsoon 2002.

Climate change and agriculture

Based on some of the past experiences indicated above, impact of climate change on agriculture will be one of the major deciding factors influencing the future food security of mankind on the earth. Agriculture is not only sensitive to climate change but also one of the major drivers for climate change. Understanding the weather changes over a period of time and adjusting the management practices towards achieving better harvest are challenges to the growth of agricultural sector as a whole. The climate sensitivity of agriculture is uncertain, as there is regional variation in rainfall, temperature, crops and cropping systems, soils and management practices. The inter-annual variations in temperature and precipitation were much higher than the predicted changes in temperature and precipitation. The crop losses may increase if the predicted climate change increases the climate variability. 6 Different crops respond differently as the global warming will have a complex impact. The tropics are more dependent on agriculture as 75% of world population lives in tropics and two thirds of these people's main occupation is agriculture. With low levels of technology, wide range of pests, diseases and weeds, land degradation, unequal land distribution and rapid population growth, any impact on tropical agriculture will affect their livelihood. Rice, wheat, maize, sorghum, soybean and barley are the six major crops in the world grown in 40% cropped area, and contribute to 55% of non-meat calories and over 70% of animal feed (FAO, 2006). Consequently, any effect on these crops would adversely affect the food security.

Main projections for climate change at global level:

The projections of future climate patterns are largely based on computer-based models of the climate system that incorporate the important factors and processes of the atmosphere and the oceans, including the expected growth in greenhouse gases from socio-economic scenarios for the coming decades. The IPCC has examined the published results from many different models and on the basis of the evidence has estimated that by 2100-

- » The global average surface warming (surface air temperature change) will increase by 1.1 - 6.4 °C.
- » The sea level will rise between 18 and 59 cm.
- » The oceans will become more acidic.
- » It is very likely that hot extremes, heat waves and heavy precipitation events will continue to become more frequent.
- » It is very likely that there will be more precipitation at higher latitudes and it is likely that there will be less precipitation in most subtropical land areas.
- » It is likely that tropical cyclones (typhoons and hurricanes) will become more intense, with larger peak wind speeds and heavier precipitation associated with ongoing increases of tropical sea surface temperatures.

Likely effects of climate change on key sectors at global level: The IPCC Fourth Assessment Report of the Working Group II: Impacts, Adaptation and Vulnerability describe the likely effects of climate change, including from increases in extreme events. The effects on key sectors, in the absence of countermeasures, are summarized as follows.

Water: Drought affected areas are likely to be more widely distributed. Heavier precipitation events are very likely to increase in frequency leading to higher flood risks. By mid-century, water availability is likely to decrease in mid-latitudes, in the dry tropics and in other regions supplied by melted water from mountain ranges. More than one sixth of the world's population is currently dependent on melt water from mountain ranges.

Food: While some mid latitude and high latitude areas will initially benefit from higher agricultural production, for many others at lower latitudes, especially in seasonally dry and tropical regions, the increases in temperature and the frequency of droughts and floods are likely to affect crop production negatively, which could increase the number of people at risk from hunger and increased levels of displacement and migration.

Industry, settlement and society: The most vulnerable industries, settlements and societies are generally those located in coastal areas and river flood plains, and those whose economies are closely linked with climate sensitive resources. This applies particularly to locations already prone to extreme weather events and especially to areas undergoing rapid urbanization. Where extreme weather events become more intense or more frequent, the economic and social costs of those events will increase.

Health: The projected changes in climate are likely to alter the health status of millions of people, including increased deaths, disease and injury due to heat waves, floods, storms, fires and droughts. Increased malnutrition, diarrhea disease and malaria in some areas will increase vulnerability to extreme public health, and development goals will be threatened by long term damage to health systems from disasters.

Projected impact on Asia

- Asia-Pacific region may experience the worst effect on rice and wheat yields worldwide, and decreased yields could threaten the food security of 1.6 billion people in South Asia. 8
- The crop model indicates that in South Asia, average yields in 2050 for crops will decline from 2000 levels by about 50 per cent for wheat, 17 percent for rice, and about 6 percent for maize because of climate change.
- In East Asia and the Pacific, yields in 2050 for crops will decline from 2000 levels by 20 percent for rice, 13 percent for soybean, 16 percent for wheat and 4 percent for maize because of climate change.
- With climate change, average calorie availability in Asia in 2050 is expected to be about 15 percent lower and cereal consumption is projected to decline by as much as 24 percent compared to a no-climate change scenario.
- In a no-climate change scenario, the number of malnourished children in South Asia would fall from 76 to 52 million between 2000 and 2050, and from 24 to 10 million in East Asia and the Pacific.
- Climate change will erase some of this progress, causing the number of malnourished children in 2050 to rise to 59 million in South Asia and to 14 million in East Asia and the Pacific, increasing the total number of malnourished children in Asia by about 11 million.
- To counteract the effects of climate change on nutrition, South Asia requires additional annual investments of 1.5 billion USD in rural development, and

East Asia and the Pacific require almost 1 million USD more. Over half of these investments in both regions must be for irrigation expansion.

» The Asian countries most vulnerable to climate change are Afghanistan, Bangladesh, Cambodia, India, Lao PDR, Myanmar, and Nepal.

» Afghanistan, Bangladesh, India, and Nepal are particularly vulnerable to declining crop yields due to glacial melting, floods, droughts, and erratic rainfall, among other factors.

» Asia is the most disaster-afflicted region in the world, accounting for about 89 percent of people affected by disasters worldwide.

Observed changes in climate and weather events in India

Surface temperature

At the national level, increase of 0.4° C has been observed in surface air temperatures over the past century. A warming trend has been observed along the west coast, in central India, the interior peninsula, and northeastern India. 9 However, cooling trends have been observed in northwest India and parts of south India.

Rainfall

While the observed monsoon rainfall at the All India level does not show any significant trend, regional monsoon variations have been recorded. A trend of increasing monsoon seasonal rainfall has been found along the west coast, northern Andhra Pradesh, and north-western India (+10% to +12% of the normal over the last 100 years) while a trend of decreasing monsoon seasonal rainfall has been observed over eastern Madhya Pradesh, north-eastern India, and some parts of Gujarat and Kerala (-6% to - 8% of the normal over the last 100 years)

Extreme weather events

Trends of Extreme Weather Events observed in multi-decadal periods of more frequent droughts followed by less severe droughts. There has been an overall increasing trend in severe storm incidence along the coast at the rate of 0.011 events per year. While the states of West Bengal and Gujarat have reported increasing trends, a decline has been observed in Orissa. Scientists, while analysing a daily rainfall data set, have shown (i) a rising trend in the frequency of heavy rain events, and (ii) a significant decrease in the frequency of moderate events over central India from 1951 to 2000.

Rise in sea level

Using the records of coastal tide gauges in the north Indian Ocean for more than 40 years, Scientists have estimated that sea level rise was between 1.06-1.75 mm per year. These rates are consistent with 1-2 mm per year global sea level rise estimates of the IPCC.

Indian Summer Monsoon (ISM) intensity is projected to increase in the beginning of 2040 and by 10% by 2100.

Impacts on Himalayan glaciers

The Himalayas possess one of the largest resources of snow and ice and its glaciers form a source of water for the perennial rivers such as the Indus, the Ganga, and the Brahmaputra. Glacial melt may impact their long-term lean-season 10 flows, with adverse impacts on the economy in terms of water availability and hydropower generation. The available monitoring data on Himalayan glaciers indicates that while recession of some glaciers has occurred in some Himalayan regions in recent years, the trend is not consistent across the entire mountain chain.

Some projections of climate change over India for the 21st century

Some modelling and other studies have projected the following changes due to increase in atmospheric GHG concentrations arising from increased global anthropogenic emissions:

Annual mean surface temperature

The simulation studies by Indian Institute of Tropical Meteorology (IITM), Pune, estimated that annual mean surface temperature is expected to raise by the end of century, ranges from 3 to 5 °C with warming more pronounced in the northern parts of India.

Impacts on water resources

Changes in key climate variables, namely temperature, precipitation and humidity, may have significant long-term implications for the quality and quantity of water. River systems of the Brahmaputra, the Ganga, and the Indus, which benefit from melting snow in the lean season, are likely to be particularly affected by the decrease in snow cover. A decline in total run-off for all river basins, except Narmada and Tapti, is projected in India's NATCOM I. A decline in run-off by more than twothirds is also anticipated for Sabarmati and Luni basins. Due to sea level rise, the fresh water sources near the coastal regions will suffer salt intrusion.

Impacts on agriculture and food production

Food production in India is sensitive to climate changes such as variability in monsoon rainfall and temperature changes within a season. Studies by Indian Agricultural Research Institute (IARI) and others indicate greater expected loss in the Rabi crop. Every 1°C rise in temperature reduces wheat production by 4-5 Million 11 Tonnes. Small changes in temperature and rainfall have significant effects on the quality of fruits, vegetables, tea, coffee, aromatic and medicinal plants, and basmati rice. Pathogens and insect populations are strongly dependent upon temperature and humidity, and changes in these parameters may change their population dynamics. Other impacts on agricultural and related sectors include lower yields from dairy cattle and decline in fish breeding, migration, and harvests. Global reports indicate a loss of 10-40% in crop production by 2100.

Indian climate is dominated by the southwest monsoon, which brings most of the region's precipitation. It is critical for the availability of drinking water and irrigation for agriculture. Agricultural productivity is sensitive to two broad classes of climate-induced effects (1) direct effects from changes in temperature, precipitation or carbon dioxide concentrations, and (2) indirect effects through changes in soil moisture and the distribution and frequency of infestation by pests and diseases. Rice and wheat yields could decline considerably with climatic changes (IPCC 1996; 2001). However, the vulnerability of agricultural production to climate change depends not only on the physiological response of the affected plant, but also on the ability of the affected socio-economic systems of production to cope with changes in yield, as well as with changes in the frequency of droughts or floods. The adaptability of farmers in India is severely restricted by the heavy reliance on natural factors and the lack of complementary inputs and institutional support systems. The loss in net revenue at the farm level is estimated to range between 9% and 25% for a temperature rise of 2 °C to 3.5 °C. Scientists also estimated that a 2°C rise in mean temperature and a 7% increase in mean precipitation would reduce net revenues by 12.3% for the country as a whole. Agriculture in the coastal regions of Gujarat, Maharashtra, and Karnataka is found to be the most negatively affected. Small losses are also indicated for the major food-grain producing regions of Punjab, Haryana, and western Uttar Pradesh. On the other hand, West Bengal, Orissa, and Andhra Pradesh are predicted to benefit to a small extent from warming.

Impacts on health

Changes in climate may alter the distribution of important vector species (for example, malarial mosquitoes) and may increase the spread of such diseases to new areas. If there is an increase of 3.8 °C in temperature and a 7% increase in relative humidity, the transmission windows i.e., months during which mosquitoes are active, 12 will be open for all 12 months in 9 states in India. The transmission windows in Jammu and Kashmir and in Rajasthan may increase by 3-5 months. However, in Orissa and some southern states, a further increase in temperature is likely to shorten the transmission window by 2-3 months.

Impacts on forests

Climate projections indicate that the country is likely to experience shift in forest types, with consequent changes in forests produce, and, in turn, livelihoods based on those products. Correspondingly, the associated biodiversity is likely to be adversely impacted.

Impacts on coastal areas

A mean Sea Level Rise (SLR) of 15-38 cm is projected along India's coast by the mid 21st century and of 46-59 cm by 2100. In addition, a projected increase in the intensity of tropical cyclones poses a threat to the heavily populated coastal zones in the country (NATCOM, 2004).

Impacts on biodiversity

The Intergovernmental Panel on Climate Change has projected that global average temperature increase during 21st century will range from 1.4o to 4o Celsius. Research by the Consultative Group on International Agricultural Research based on distribution models of wild relatives of three staple crops of the poor i.e. Peanuts, 13 cowpea and potato suggests that 16-22 per cent of wild species will be threatened by extinction by 2055. Loss of genetic diversity can have serious long-term consequences globally.

Vulnerability to extreme events

Heavily populated regions such as coastal areas are exposed to climatic events such as cyclones, floods, drought, and large declines in sown areas in arid and semiarid zones occur during climate extremes. Large areas in Rajasthan, Andhra Pradesh,

Gujarat, and Maharashtra and comparatively small areas in Karnataka, Orissa, Madhya Pradesh, Tamil Nadu, Bihar, West Bengal, and Uttar Pradesh are frequented by drought. About 40 million hectares of land is flood-prone, including most of the river basins in the north and the northeastern belt, affecting about 30 million people on an average each year. Such vulnerable regions may be particularly impacted by climate change.

Impacts on pests

Some of the most dramatic effects of climate change on pests and diseases are likely to be seen among arthropod insects like mosquitoes, midges, ticks, fleas and sand flies, and the viruses they carry. With changes in temperature and humidity levels, the populations of these insects may expand their geographic range, and expose animals and humans to diseases to which they have no natural immunity. Plant pests, which include insects, pathogens and weeds, continue to be one of the biggest constraints to food and agricultural production. Fruit flies, for instance, cause extensive damage to fruits and vegetables production. Controlling such pests often requires the use of pesticides, which can have serious side effects on human health and the environment. Climate change may also play a role in food safety. A growing number of pests and diseases could lead to higher and even unsafe levels of pesticide residue and veterinary drugs in local food supplies. And changes in rainfall, temperature and relative humidity can readily contaminate foods like groundnuts, wheat, maize, rice and coffee with fungi that produce potentially fatal mycotoxins.

i. **Effect of rising temperature:** In temperate climate, increased temperature could increase insect population. Rising temperature may affect insect survival, development, geographic range and population size. It may affect insect physiology. Under such situation some insects take several years to complete life cycle (Cicadas, Arctic moths) and some insects develop quickly at certain temperature range based on degree days (cabbage maggot, onion maggot, European corn borer, Colorado potato beetle, aphids, diamond back moth). Therefore, crop damage increase. Migratory pests may migrate earlier. Natural enemy-host relationship may affect resulting into reduced parasitism. Rising temperature may change gender ratios of insects such as thrips. The population of insects will increase due to lower winter mortality of insects as a result of warmer winter. Higher temperature may tend to shift crops geographically and hence its pests to higher altitudes. From fossil records it is understood that diversity of insect species and intensity of feeding increase with increase in temperature. Increased temperature could decrease insect population (aphids) in some crops, which cannot be grown in higher temperature. The same 14 condition may be conducive for increased activity of natural enemies of that pest further reducing its population.

ii. **Effect of Precipitation:** Rain drops physically dislodge the insects from their hosts such as leafhoppers, plant hoppers, thrips, cut worms etc. while others drown to death e.g. mealy bugs, pupae of fruit fly, Helicoverpa, Spodoptera, Etiella , rice stem borers etc. Flooding is used as a control measure for termites and stem borers too. Heavy rainfall causes pest epizootics by fungal pathogens (sugarcane pyrilla). It is anticipated that cutworm infestation will be more in future because they are sensitive to flooding and summer rainfall, which will increase in future.

iii. **III) Effect of rising CO_2 level:** Carbon dioxide is a perfect example of a change that could have both positive and negative effects. Carbon dioxide is expected to have positive physiological effects through increased photosynthesis. The impact is higher on C3 crops such as wheat and rice than on C4 plants like maize and grasses. The direct effects of changes in CO_2 concentration will be through changes in temperature, precipitation and radiation. However, indirect effects will bring changes in soil moisture and infestation by pests and diseases because of rising temperature and relative humidity. Such indirect effects through the increase in temperature will reduce crop duration, increase crop respiration rates, evapo-transpiration, decrease fertilizer use efficiency and enhance pest infestation. There is general consensus that the yield of main season (Kharif) crop will increase due to the effect of higher CO_2 levels. However, large yield decreases are predicted for the rabi crops because of increased temperatures. The rising CO_2 level in atmosphere has indirect impact on insect population. Soybean crop in higher CO_2 concentration had 57% more insect damage (Japanese beetle, Leafhopper, Root worm, Mexican bean beetle) than earlier. It causes increase in level of simple sugars in the leaves that stimulates more feeding by insects. Increased C/N ratio in plant tissue due to increased CO_2 level may slow insect development and increase life stages of insect pests vulnerable to attack by parasitoids. At our current rate of green house emissions, several of the main pests that target corn will increase in number and expand their range by the end of 21st century.

iv. **Effect on insecticide Use Efficiency:** Warmer temperature requires more number of insecticide applications (i.e., three, more than normal) for controlling corn 15 pests. Entomologists predict more generation of insects in warm climate that necessitates more number of insecticide applications. It will increase cost of protection and environmental pollution. Synthetic pyrethroids and naturalite spinosad will be less effective in higher temperature. Therefore, it is advisable for the farmers not to use insecticides with similar mode of action frequently to avoid development of resistance in case of more number of applications. Cultural management practices e.g. early planting may not be helpful because of early emergence of pests due to warmness.

v. **Effect on natural pest control:** Global warming is expected to make regional climates more varied and unpredictable which could affect relationship between insects and their natural enemies. In years of most variable rainfall, the caterpillars have significantly less number of parasitoids. This could be because the parasitoids use cues e.g. change in local climate to determine the best time for laying eggs. Unpredictable rains might disrupt the parasitoids ability to track their caterpillar hosts. The wasps use start of the rain as cues to hatch out of their cocoons and look for a caterpillar to lay their eggs. If the rains are late, they emerge late and may not find larval stage of host resulting in reduced natural pest control. Due to changes in climate, the frequency of occurrence of droughts, heat waves, windstorms and floods etc. will increase disrupting the natural ecosystems.

vi. **Effect on Forest insect pests:** It is difficult to predict impacts of climate change on forest insect pests because of complexity of interactions between insects and trees. Population of green spruce aphid (Elatobium abietinum) will increase due to global warming. The spruce bark beetle (Dendroctonus micans) will increase due to warming because its predator Rhizophagus grandis is benefited by temperature rise. The Asian long horn beetle (Anoplophora glabripennis) population will increase in warmer coastal areas that attack street plantations. In general, it is assumed that many forest insect pests will increase as a result of climate change. At the same time, it is likely that pests' natural enemies will benefit. So it is unclear to some extent as to what the overall effect of global climate change will be on forest insect pests.

Impact of climate change on diseases

Any direct yield gains caused by increased CO_2 or climate change could be offset partly or entirely by losses caused by phytophagous insects, plant pathogens and weeds. It is, therefore, important to consider these biotic constraints on crop yields under climate change.

i. **Impacts on plant pathosystems:** Climate change has the potential to modify host physiology and resistance and to alter stages and rates of development of the pathogen. The most likely impacts would be shifts in the geographical distribution of host and pathogen, changes in the physiology of host-pathogen interactions and changes in crop loss. Another important impact may be through changes in the efficacy of control strategies.

ii. **Geographical distribution of host and pathogen:** New disease complexes may arise and some diseases may cease to be economically important if warming causes a pole ward shift of agro climatic zones and host plants migrate into new regions. Pathogens would follow the migrating hosts and may infect remnant vegetation of natural plant communities not previously exposed to the often more aggressive strains from agricultural crops. The mechanism of pathogen dispersal, suitability of the environment for dispersal, survival between seasons, and any change in host physiology and ecology in the new environment will largely determine how quickly pathogens become established in a new region. Changes may occur in the type, amount and relative importance of pathogens and affect the spectrum of diseases affecting a particular crop. This would be more pronounced for pathogens with alternate hosts. Plants growing in marginal climate could experience chronic stress that would predispose them to insect and disease outbreaks. Warming and other changes could also make plants more vulnerable to damage from pathogens that are currently not important because of unfavorable climate.

iii. **Physiology of host-pathogen interactions – elevated CO_2:** Increases in leaf area and duration, leaf thickness, branching, tillering, stem and root length and dry weight are well known effects of increased CO_2 on many plants. Scientists have suggested that elevated CO_2 would increase canopy size and density, resulting in a greater biomass of high nutritional quality. When combined with increased canopy humidity, this is likely to promote foliar diseases such as rusts, powdery 17 mildews, leaf spots and blights. The decomposition of plant litter is an important factor in nutrient cycling and in the saprophytic survival of many pathogens. Increased C:N ratio of litter is a consequence of plant growth under elevated CO_2. Scientists have indicated that decomposition of high- CO_2 litter occurs at a slower rate. Increased plant biomass, slower decomposition of litter and higher winter temperature could increase pathogen survival on over wintering crop residues and increase the amount of initial inoculum available to infect subsequent crops. Hostpathogen interactions in selected fungal pathosystems, two important trends have emerged on the effects of elevated CO_2. First, the initial establishment of the pathogen may be delayed because of modifications in pathogen aggressiveness and/or host susceptibility. The second important finding has been an increase in the fecundity of pathogens under elevated CO_2.

iv. **Elevated temperature:** Increases in temperature can modify host physiology and resistance. Considerable information is available on heat-induced susceptibility and temperature-sensitive genes. In contrast, lignification of cell walls increased in forage species at higher temperatures to enhance resistance to fungal pathogens. Impacts would, therefore, depend on the nature of the host-pathogen interactions and the mechanism of resistance. Agricultural crops and plants in natural communities may harbor pathogens as symptomless carriers, and disease may develop if plants are stressed in a warmer climate. Host stress is an especially important factor in decline of various forest species.

Crop loss: At elevated CO_2, increased partitioning of assimilates to roots occurs consistently in crops such as carrot, sugar beet and radish. If more carbon is stored in roots, losses from soil borne diseases of root crops may be reduced under climate change. In contrast, for foliar diseases favored by high temperature and humidity, increases in temperature and precipitation under climate change may result in increased crop loss. The effects of enlarged plant canopies from elevated CO_2 could further increase crop losses from foliar pathogens.

Research finding of ICAR on climate change: To meet the challenges as posed by climate change on the agricultural system, ICAR has accorded high priority in understanding the impacts of climate change and developing adaptation and 18 mitigation strategies through its network research programs. Some of the research findings are as follows:

» Significant negative rainfall trends were observed in the Eastern parts of Madhya Pradesh, Chhattisgarh and parts of Bihar, Uttar Pradesh, parts of northwest and NE India and also a small pocket in Tamil Nadu.

» Significant increase in rainfall has also been noticed in Jammu and Kashmir and in some parts of southern peninsular. The maximum and minimum temperature (1960-2003) analysis for northwest region of India showed that the minimum temperature is increasing at annual, *kharif* and rabi season time scales. The rate of increase of minimum temperature during rabi is much higher than during *kharif*. The maximum temperature showed increasing trend in annual, *kharif* and rabi time scales but very sharp rise was observed from the year 2000 onwards.

» It was observed from the experiments on impact of high temperature on pollen sterility and germination in rice that maximum temperature above 35°C and minimum temperature 23°C at flowering stage increased the pollen sterility in two normal and three basmati varieties of rice and the effect is more profound in basmati cultivars.

» Biological yields were reduced drastically with elevated ambient temperature in tunnel experiments. The degrees of reduction in grain yield enhanced with rise in ambient temperature at 1, 2 and 3°C. The reduction of grain yield by 60, 64 and 70 percent in Pusa Sugandh-2 and 45, 52, 54 percent in Pusa 44 variety which was mainly attributed to maximum reduction in number of panicles/m2 followed by the number of panicles/m2 and 1000 grain weight.

» High thermal stress during post-flowering duration manifested 18, 60 and 12 percent reduction in economic yield of wheat, mustard and potato, respectively.

» Coconut yields were not affected with the increase of maximum temperature up to 44°C but above that reduced the yield.

» The growth and yield response of castor crop at first and second germination levels showed positive response to enhanced CO_2.

» The reproductive phase (days to flowering) and maturity phase shortened by 5 and 15 days in early and late sown varieties of wheat at Palm Valley of Himachal Pradesh. One to ten days shortening of reproductive phase in rice was observed in Palampur region.

» Increasing temperature above 1°C in the Himalayan region is adversely affecting the yield of apple. In Himachal Pradesh, rainfall at low (1100 m) and mid (1800-2000 m) elevation has declined and erratic. At higher ranges (2600-2700 m) snowfall has declined from 10 feet (40 years back) to 1-2 feet in the recent years.

» Deodar, Kail and Kharsu are drying and dying (yellowing) at (1700-2300 m) elevation, whereas at higher elevation (2500 m) Insect attack in oak was observed in Shimla region.

» A rise in 2-6°C temperature impacts the growth, puberty and maturity of cross breeds and buffaloes. The time to attain puberty prolongs from 1 to 2 weeks because of the slow growth rates at higher temperature.

» Milk production in Holstein Friesian cross breed cows was affected due to rise in maximum and minimum temperatures above 22°C. Decrease of milk production in Murra buffaloes was also observed with increase in temperature above 2°C. The extreme events like heat wave (> 4°C) and cold wave (< 3°C) reduced the milk yield by 10-30 percent in first lactation and 5-20 percent in second and third lactations in cattle and buffaloes.

- Total methane emission due to enteric fermentation and manure management of 485 million heads of livestock was worked out at 9.36 Tg/annum for the year 2006. It was 9.32 Tg/annum in the year 2003.
- Reduction of methane emission in cattle (cross breed steers) was achieved by modifying the diet by supplementing fenugreek seeds (Trigonella foerum).
- Trends in Surface Sea Temperature (SST) showed significant increase at the rate of 0.045 °C per decade along the southwest, northwest and northeast coasts whereas the rate of increase of 0.095 °C per decade was observed along the southeast coast.
- The oil sardine fish once restricted to southwest coast along 8°N to 12°N was extended along the other coastal areas and also extended into Bay of Bengal up to Orissa and West Bengal coast due to congenial environment prevailed with the increasing SSTs.
- A shift in lower stretch fish species like Puntim ticto, Xenentodon cancila, Mystus vittatus and Glossogobius giuris, etc. to the cold water rithron zone of the river Ganga at Haridwar due to rise in average temperature condition of the river from 17.5 °C to 25.5 °C.

Some current actions for adaptation and mitigation in India

Adaptation, in the context of climate change, comprises the measures taken to minimize the adverse impacts of climate change, e.g. relocating the communities living close to the sea shore, for instance, to cope with the rising sea level or switching to crops that can withstand higher temperatures. Mitigation comprises measures to reduce the emissions of greenhouse gases that cause climate change in the first place, e.g. by switching to renewable sources of energy such as solar energy or wind energy or nuclear energy instead of burning fossil fuel in thermal power stations.

Current government expenditure in India on adaptation to climate variability exceeds 2.6% of the GDP, with agriculture, water resources, health and sanitation, forests, coastal-zone infrastructure and extreme weather events, being specific areas of concern.

Programmes

Crop improvement: Programmes address measures such as development of aridland crops and pest management as well as capacity building of extension workers and NGOs to support better vulnerability-reducing practices.

Drought proofing: The current programmes seek to minimize the adverse effects of drought on production of crops and livestock, and on productivity of land, water and human resources, so as to ultimately lead to drought proofing of the affected areas. They also aim to promote overall economic development and improve the socio-economic conditions of the resource poor and disadvantaged.

Forestry: India has a strong and rapidly growing afforestation programme. The afforestation process was accelerated by the enactment of the Forest Conservation Act of 1980, which aimed at stopping the clearing and degradation of forests through a strict, centralized control of the rights to use forestland and mandatory requirements of compensatory afforestation in case of any diversion of forestland for any non-forestry purpose. In addition, an aggressive afforestation and sustainable forest management programme resulted in annual reforestation of 1.78 mha during 1985-1997, and is currently 1.1 mha annually. Due to this, the carbon stocks in Indian forests have increased over the last 20 years to 9 -10 gigatons of carbon (GtC) during 1986 to 2005.

Water: The National Water Policy (2002) stresses that non-conventional methods for utilization of water, including inter-basin transfers, artificial recharge of groundwater, and desalination of brackish or sea water, as well as traditional water conservation practices like rainwater harvesting, including roof-top rainwater harvesting, should be practiced to increase the utilizable water resources. Many states now have mandatory water harvesting programmes in several cities.

Coastal regions: In coastal regions, restrictions have been imposed in the area between 200 m and 500 m of the HTL (High Tide Line) while special restrictions have been imposed in the area up to 200 m to protect the sensitive coastal ecosystems and prevent their exploitation. This, simultaneously, addresses the concerns of the coastal population and their livelihood. Some specific measures taken 22 in this regard include construction of coastal protection infrastructure and cyclone shelters, as well as plantation of coastal forests and mangroves.

Risk financing: Two risk-financing programmes support adaptation to climate impacts. The Crop Insurance Scheme supports the insurance of farmers against climate risks, and the Credit Support Mechanism facilitates the extension of credit to farmers, especially for crop failure due to climate variability.

Disaster management: The National Disaster Management programme provides grants-in-aid to victims of weather related disasters and manages disaster relief operations. It also supports proactive disaster prevention programmes, including dissemination of information and training of disaster management staff.

India's policy structure relevant to GHG mitigation

India has in place a detailed policy, regulatory and legislative structure that relates strongly to GHG mitigation: The Integrated Energy Policy was adopted in 2006. Some of its key provisions are:

- » Promotion of energy efficiency in all sectors
- » Emphasis on mass transport
- » Emphasis on renewable including biofuels plantations
- » Accelerated development of nuclear and hydropower for clean energy
- » Focused R & D on several clean energy related technologies

The experience gained so far enables India to embark on an even more proactive approach through National Action Plan on Climate Change (NAPCC). NAPCC identifies measures that promote our development objectives while also yielding co-benefits for addressing climate change effectively. It outlines a number of steps to simultaneously advance India's development and climate change related objectives of adaptation and mitigation. The following eight National Missions form the core of the National Action Plan, representing multi-pronged, long-term and integrated strategies for achieving key goals in the context of climate change:

1. National Solar Mission
2. National Mission for Enhanced Energy Efficiency
3. National Mission on Sustainable Habitat
4. National Water Mission
5. National Mission for Sustaining the Himalayan Ecosystem
6. National Mission for a "Green India" vii. National Mission for Sustainable Agriculture
7. National Mission on Strategic Knowledge for Climate Change

Coping options for farmers awareness on climate change

Farmers need to be sensitized on climate variability, climate change, its impact on crop production, and coping options.

Agromet advisories: The farming community is provided with Agromet advisories. Its bulletins are prepared taking into account the prevailing weather, soil and crop condition and weather prediction. Measures / practices / suggestions to be taken in view of weather forecast to minimize the losses and optimize inputs (Land preparation, selection of crop & cultivars, Date of sowing, Date of harvesting, Irrigation scheduling, Pesticides & Fertilizer application, Extreme weather events, etc.). Agro-advisory bulletin consist of three parts (i) Weather events occurred during past week and weather forecast for five days ahead. These forecast includes weather parameters like cloud amount, rainfall, average Wind Speed, Wind Direction, RH, maximum and minimum temperature. (ii) It contains actual information on state and stage of crop growth, ongoing agricultural operations, disease and insect pest occurrence. (iii) It provides value added information on various farm activities to be taken based on weather There are 23 State Agromet Service Centers. These prepare Agromet advisory in collaboration with State Department of Agriculture on Tuesday & Friday. The Agromet advisory is disseminated through All India Radio (AIR), Print Media, Doordarshan, Website and SMS.

An insurance product based on a Weather Index:

The basic idea of weather insurance is to estimate the percentage deviation in crop output due to adverse weather conditions. Weather insurance protects against additional expenses or loss of profit from specific bad weather events.

Contingency planning: It is a plan devised for an exceptional risk, which is impractical or impossible to avoid. Ministry of Agriculture, through Indian Council of Agricultural Research (ICAR) and State Agricultural Universities (SAUs), is working on district-specific contingency plan for the agriculture and allied sectors. This includes fisheries and animal husbandry and started in March 2010 under Rashtriya Krishi Vikas Yojana (RKVY). Contingency plans for about 300 out of the total 600 odd districts were prepared and validated by experts and hosted in Department of Agriculture and Cooperation (DAC), ICAR, Centre for Research in Dry land Areas (CRIDA) website. The comprehensive district-specific document is having details on the crops and cultivation practices to be adopted in case of deficient or delay in monsoon, unseasonal rains, frosts or unusually high temperature, excessive rains etc. Each district is a scientific document for adaptation in case of eventualities.

Demonstration of climate resilient technologies to the farmers may be undertaken by the extension personnel in the areas of:

i. **Natural resource management:** Interventions on in-situ moisture conservation, rain water harvesting and recycling for supplemental irrigation, improved drainage in flood prone areas, conservation tillage, ground water recharge and water saving irrigation methods etc.

ii. **Crop production:** Introducing drought/temperature tolerant varieties, advancement of planting dates of rabi crops in areas with terminal heat stress, water saving paddy cultivation methods (SRI, aerobic, direct seeding), frost management in horticulture through fumigation, community nurseries for delayed monsoon, custom hiring centers for farm machineries for timely planting, location specific intercropping systems with high sustainable yield index etc

iii. **Livestock and fisheries:** Augmentation of fodder production during droughts/ floods, improving productivity of Common Property Resources (CPRs), promotion of improved fodder/feed storage methods, preventive vaccination, improved shelters for reducing heat/cold stress, management of fish ponds/tanks during water scarcity and flooding etc.

iv. **Institutional interventions:** Institutional interventions, either by strengthening the existing ones or initiating new ones, relating to seed 25 bank, fodder bank, custom hiring center, collective marketing, and introduction of weather index based insurance and climate literacy through a village level weather station.

v. **Extension system** has to focus more on diversifying the livelihood options, changing suitable cropping patterns to adjust to the change which is occurring in the particular location, planting more drought tolerant crops, promoting increased share of non-agricultural activities and Agro-forestry practices, identifying the traditional coping strategies, improved on - farm soil & water conservation, promoting mixed cropping pattern and making provision for access to various information sources related to weather and other advisories of climate change would minimize the risks and certainty of farmers related to climate change.

Conclusion

From the above discussion, it is clear that the occurrence of floods and droughts, heat and cold waves are common across the world due to climate change. Their adverse impact on livelihood of farmers is tremendous. It is more so in India as our economy is more dependent on Agriculture. Interestingly, weather extremes of opposite in nature like cold and heat waves and floods and droughts are noticed within the same year over the same region or in different regions and likely to increase in ensuing decades. The human and crop losses are likely to be heavy. The whole climate change is associated with increasing greenhouse gases and human induced aerosols and the imbalance between them may lead to uncertainty even in year-to-year monsoon behaviour over India. Therefore, there should be a determined effort from developed and developing countries to make industrialization environment friendly by reducing greenhouse gases pumping into the atmosphere. In the same fashion, awareness programmes on climate change and its effects on various sectors viz., agriculture, health, infrastructure, water, forestry, fisheries, land and ocean biodiversity and sea level and the role played by human interventions in climate change need to be taken up on priority basis. In the process, lifestyles of people should also be changed so as not to harm earth atmosphere continuum by pumping greenhouse gases and CFCs into the atmosphere. From the agriculture point of view, effects of extreme weather events on crops are to be documented on regional scale 26 so that it will be handy to planners in such re-occurrence events for mitigating the ill effects. Also, there is need to guide farmers on projected impact climate change and sensitise them on probable mitigation and adaptation options to minimize the risk in Agricultural sector.

Chapter - 19

Exploitation of Solar Energy in Agriculture

Energy is one of the highest overhead costs in agriculture, more so for farmers with greenhouses. Traditional sources of power, such as fossil fuels, harm the environment affecting agriculture. Development of any region is reflected in its quantum of energy consumption. With a view to keep pace with development we have to grow our energy resources at minimum 6%. The electricity problem is more severe in rural areas of India where 70% of population live and have agriculture as the main occupation. However, statistically few thousand villages are yet to be electrified, the availability of regular supply in far off places is a problem and the farmers are unable to derive benefits of electricity. The fast depleting kerosene is used for lighting and diesel oil for running agricultural machinery including pumps. Reliance on fossil fuels is unsustainable as the resource are declining. Besides people burn firewood, agricultural waste and cow dung-cake for cooking food causing irreparable damage to the eco-system. Solar energy comes from a never-ending resource. In the context of rural economy, energy is basically needed for cooking food, heating of water; lighting of houses at the domestic front while in agricultural sector energy is required for field operations, pumping water, spraying of insecticides, post-harvest activities and running of agro-and cottage industries. The situation is still worse in arid and semi arid regions where biomass is scarce and there is no hydro-electricity. Besides, farmers are unable to generate additional income due to lack of energy resources to run appropriate device in cottage industries. The arid and semi-arid part of the country receives higher radiation than rest of the country with 6.0 kWhm-2day^{-1} mean annual daily solar radiation received having 6 to 9 average sunshine hr/day. Further, it was estimated that solar energy of 1% of land area, wind power of 5% of land area and biogas (80% collection efficiency) can provide 1504 kWh year^{-1}

energy per caput in hot region while the average per caput total energy consumption of India is 1,122 kWh $year^{-1}$. In this context, renewable sources of energy like solar energy, wind power and biogas need to be harnessed for the sustainable development in general and catering the farmer requirements in particular. Therefore, adopting solar energy for agriculture alleviates the costs, and it is good for the environment.

Solar energy is applicable in various agricultural activities. Such activities include drying, cooling, and generating steam. Other applications of solar energy are pumping water for irrigation and many others.

Solar energy-powered water pumps

In most areas without access to electricity, Photovoltaic water pumping systems are effective. Additionally, these systems provide water to remote areas too. Usually, simple PV systems are designed to work when the sun is shining, providing water where needed. In these cases, solar storage batteries are unnecessary because farmers store the water in tanks or pump it directly to fields. However, larger systems can use tracking mounts, storage channels, and inverters.

Farmers who install a large-sized PV system do little maintenance because the power is reliable throughout. These systems are useful for irrigation, livestock water supply, pond aeration, and many others.

Water and space heating

Livestock and dairy operations have varying space and water heating demands. Modern-day farmers rear animals such as cattle and poultry in enclosed structures. Due to this, the needs for temperature and air quality control are vital for such buildings. To remove moisture, dust, and toxic odors, the farmer needs to replace air regularly. Also, heating such large spaces requires a considerable amount of energy. Incorporating a well-designed solar energy system can heat the air even before it gets into the building. The system can also supply additional ventilation.

Apart from heating spaces, solar is applied in agriculture to heat water. A solar water heating system reduces the costs for the farm. Mostly, in dairy farms where energy costs are high by 40%.

Crops and grain drying

This is a traditional method of utilising solar energy for drying of agricultural and animal products. Agricultural products are dried in a simple cabinet dryer which consists of a box insulated at the base, painted black on the inner side and covered with an inclined transparent sheet of glass.

At the base and top of the sides ventilation holes are provided to facilitate the flow of air over the drying material which is placed on perforated trays inside the cabinet. These perforated trays or racks are carefully designed to provide controlled exposure to solar radiations.

Solar drying, especially of fruits improves fruit quality as the sugar concentration increases on drying. Normally soft fruits are particularly vulnerable to insect attack as the sugar content increases on drying but in a fruit dryer considerable time is saved by quicker drying —minimizing gap the chances of insect attack.

The present practice of drying chilies by spreading them on the floor not only requires a lot of open space and manual labour for material handling but it becomes difficult to maintain its quality and taste unless drying is done in a controlled atmosphere. Moreover, the products being sun dried very often get spoiled due to sudden rains, dust storms or by birds. Besides, reports reveal that it is not possible to attain very low moisture content in the sun-dried chilies.

As a result, the chilies become prone to attack by fungi and bacteria. In sun-drying sometimes, the produce is over dried and its quality is lost. Solar energy operated dryer helps to overcome most of these disadvantages.

Other agricultural products commonly solar-dried are potato-chips, berseem, grains of maize and paddy, ginger, peas, pepper, cashew-nuts, timber and veneer drying and tobacco curing. Spray drying of milk and fish drying are examples of solar dried animal products.

Green house heating

A green house is a structure covered with transparent material (glass or plastic) that acts as a solar collector and utilises solar radiant energy to grow plants. It has heating, cooling and ventilating devices for controlling the temperature inside the green house.

Solar radiations can pass through the green house glazing but the thermal radiations emitted by the objects within the green house cannot escape through the glazed surface. As a result, the radiations get trapped within the green house and result in an increase in temperature. As the green house structure has a closed boundary, the air inside the greenhouse gets enriched with CO2 as there is no mixing of the greenhouse air with the ambient air. Further, there is reduced moisture loss due to restricted transpiration. All these features help to sustain plant growth throughout the day as well as during the night and all year round.

Remote supply of electricity

Electric energy or electricity can be produced directly from solar energy by means of photovoltaic cells. The photovoltaic cell is an energy conversion device which is used to convert photons of sunlight directly into electricity. It is made of semi conductors which absorb the photons received from the sun, creating free electrons with high energies.

These high energy free electrons are induced by an electric field, to flow out of the semiconductor to do useful work. This electric field in photovoltaic cells is usually provided by a p-n junction of materials which have different electrical properties. There are different fabrication techniques to enable these cells to achieve maximum efficiency. These cells are arranged in parallel or series combination to form cell modules. Some of the special features of these modules are high reliability, no expenditure on fuel, minimum cost of maintenance, long life, portability, modularity, pollution free working etc.

Photovoltaic cells have been used to operate irrigation pumps, rail road crossing warnings, navigational signals, highway emergency call systems, automatic meteorological stations etc. in areas where it is difficult to lay power lines. They are also used for weather monitoring and as portable power sources for televisions, calculators, watches, computer card readers, battery charging and in satellites etc. Besides these, photovoltaic cells are used for the energisation of pump sets for irrigation, drinking water supply and for providing electricity in rural areas i.e. street lights etc.

Cooling applications

Cooling in agriculture is still a niche market though it continues to grow. PVs can also be used to refrigerate farm yields, which prevents them from going bad. Cooling is an energy-intensive process. However, in rural areas, solar refrigeration options provide an effective clean energy option. It enables production at low costs as well as reduces spoilage leading to higher yields for farmers.

Solar cooking

A variety of fuel like coal, kerosene, cooking gas, firewood, dung cakes and agricultural wastes are used for cooking purposes. Due to the energy crisis, supply of these fuels are either deteriorating (wood, coal, kerosene, cooking gas) or are too precious to be wasted for cooking purposes (cow dung can be better used as manure for improving soil fertility). This necessitated the use of solar energy for cooking purposes and the development of solar cookers. A simple solar cooker is the flat plate box type solar cooker.

It consists of a well insulated metal or wooden box which is blackened from the inner side. The solar radiations entering the box are of short wavelength. As higher wavelength radiations are unable to pass through the glass covers, the re-radiation from the blackened interior to outside the box through the two glass covers is minimised, thereby minimising the heat loss.

The heat loss due to convection is minimised by making the box airtight. This is achieved by providing a rubber strip between the upper lid and the box for minimising the heat loss due to conduction, the space between the blackened tray and outer cover of the box is filled with an insulting material like glass wool, saw-dust, paddy husk etc.

When placed in sunlight, the solar rays penetrate the glass covers and are absorbed by the blackened surface thereby resulting in an increase in temperature inside the box. Cooking pots blackened from outside are placed in the solar box.

The uncooked food gets cooked with the heat energy produced due to increased temperature of the solar box. Collector area of such a solar cooker can be increased by providing a plane reflector mirror. When this reflector is adjusted to reflect the sun rays into the box, then a 15°C to 25°C rise in temperature is achieved inside the cooker box.

The solar cooker requires neither fuel nor attention while cooking food and there is no pollution, no charring or overflowing of food and the most important advantage is that nutritional value of the cooked food is very high as the vitamins and natural tastes of the food are not destroyed. Maintenance cost of the solar cooker is negligible.

The main disadvantage of the solar cooker is that the food cannot be cooked at night, during cloudy days or at short notice. Cooking takes comparatively more time and chapattis cannot be cooked in a solar cooker.

Solar-distillation

In arid / semi and or coastal areas there is scarcity of potable water. The abundant sunlight in these areas can be used for converting saline water into potable distilled water by the method of solar distillation. In this method, solar radiation is admitted through a transparent air tight glass cover into a shallow blackened basin containing saline water.

Solar radiation passes through the covers and is absorbed and converted into heat in the blackened surface causing the water to evaporate from the brine (impure saline water). The vapors produced get condensed to form purified water in the cool interior of the roof.

The condensed water flows down the sloping roof and is collected in the troughs placed at the bottom and from there into a water storage tank to supply potable distilled water in areas of scarcity, in colleges, school science laboratories, defense labs, petrol pumps, hospitals and pharmaceutical industries. Per liter distilled water cost obtained by this system is cheaper than distilled water obtained by other electrical energy-based processes.

Conclusion

Agriculture significantly suffers from unreliable power. Some farmers have faced severe losses as a result. Solar energy for agriculture is relatively cheaper than traditional sources of electricity. Also, solar energy used for agricultural activities can be instrumental in solving drought-related issues. In areas with water scarcity, using solar energy, especially for pumping water or irrigation, can help mitigate the problem. Farmers can also experience technological advancement as the industry can become more receptive to innovation.

Solar also solves the issue of interrupted power supply. Solar power, coupled with an effective energy storage system, guarantees the farmer readily available electricity. The energy is undoubtedly one of the most important resources to be adopted in agriculture. Lastly, farmers are the most affected group when it comes to global warming. Therefore, they need to be at the center of adopting sustainable energy growth. Solar energy offers limitless opportunities in farming.

References

Altieri, M.A., 1994, Biodiversity and pest management in agro-ecosystems. Hayworth Press,New York.

Altieri, M.A., Letourneaour D.K. and Davis, J.R. 1995, Developing sustainable agro-ecosystems.*Bio-Science.*, 33: 45-49.

Altieri, M.A. and Nicholl, C.I. 1999, Biodiversity, ecosystem function and insect pest management in agricultural systems. In W.W. Collins and C.O. Qualset (eds.) Biodiversity in Agro-ecosystems. CRC Press, Boca Raton.

Balasubramanian, A. 2013, Agro-Ecological zones of India. https://www.researchgate.net/ publication/314206350.

Baker, J. T. and Allen, L. H., 1993, Contrasting crop species responses to CO_2 and temperature: rice, soybean and citrus. *Vegetatio*, **104**(1): 239-260.

Balasubramanian, A. 2019, Introduction to ecology. https://www.researchgate.net/ publication/ 335715336.

Barrios, E., Gemmill-Herren, B., Bicksler, A., Siliprandi, E., Brathwaite, R., Moller, S., Batello, C. and Tittonell, P., 2020. The 10 Elements of Agroecology: enabling transitions towards sustainable agriculture and food systems through visual narratives. *Ecosystems and People*, *16*(1), pp.230-247.

Bourne, R., 1934. Some ecological conceptions. *Empire Forestry Journal*, *13*(1), pp.15-30.

Bollero, G. A., Bullock, D. G. and Hollinger, S. E., 1996, Soil temperature and planting date effects on corn yield, leaf area, and plant development. *Agron. J.*, **88**(3): 385-390.

Botella, M.A., Cruz, C., Martins-Louçao, M.A. and Cerdá, A., 1993, Nitrate reductase activity in wheat seedlings as affected by NO3-/NH4+ ratio and salinity. *J. plant physiology*, **142**(5): 531-536.

Chaer G.M., Resende A.S., Campello E.F.C., De-Faria S.M. and Boddey R.M. (2011). Nitrogen-fixing legume tree species for the reclamation of severely degraded lands in Brazil. *Tree Physiol.*, **31:** 139-149.

Charles Elton, 1927, *Animal ecology*. The Macmillan Company, New York.

Chaudhry, F.N. and Malik, M.F., 2017. Factors affecting water pollution: a review. *J Ecosyst Ecography*, 7(225), pp.1-30.

Chauhan, A. 2020. *Environmental Pollution and Management*. I.K. International Publishing House Pvt. Limited.

Christian, R. R., 2005, Concepts of ecosystem, level and scale. Ecology–Vol. I, *Encyclopedia of Life Support Systems*. Pp1-5.

Connor, D. J., Loomis, R. S. and Cassman, K. G., 2011. *Crop ecology: productivity and management in agricultural systems*. 2nd Edn. Cambridge University Press.

Dagar J.C., Singh G. and Singh N.T. (2001). Evaluation of forest and fruit trees used for rehabilitation of semiarid alkali-sodic soils in India. *Arid Land Res. Manag.*, **15**(2)**:**115-133.

Dana, L. D. and Bauder, J.W., 2011. A general essay on bioremediation of contaminated soil. *Montana State University, Bozeman, Mont, USA*.

Dobzhansky, T., 1950. *The science of ecology today*. Quarterly review of biology. Pp. 273. In: Principles of Animal Ecology by WC Allee, Alfred E. Emerson, Thomas Park, Orlando Park, and Karl P. Schmidt. Philadelphia: WB Saunders.

Douglas P. Reagan, 2006, An ecological basis for integrated environmental management, human and ecological risk assessment. *An International Journal*, 12:5, 819-833.

Drake, B. G. and Leadley, P. W., 1991, Canopy photosynthesis of crops and native plant communities exposed to long-term elevated CO_2. *Plant Cell Environ.*, **14**(8): 853-860.

Dutta R.K. and Agrawal M. (2003). Restoration of opencast coal mine spoil by planting exotic tree species: A case study in dry tropical region. *Ecol. Eng.*, **21:** 143-151.

Erach Bharucha, 2004, *Text book of Environmental Studies* for undergraduate courses. University Grant Commission.

FAO, 2018, The 10 elements of agroecology guiding the transition to sustainable food and agricultural systems. www.fao.org.

FINCH, C.V. AND C.W. SHARP, 1976, Cover crops in california orchards and vineyards. USDA Soil Conservation Service, Washington, D.C.

GLIESSMAN, S.R, 1998, Agro-ecology: ecological processes in sustainable agriculture. Ann Arbor Press, Michigan.

Gao, L., Xu, H., Bi, H., *et al.*, 2013, Intercropping competition between apple trees and crops in agroforestry systems on the loess plateau of China. PLOS ONE 8(7): e70739. https://doi.org/10.1371/journal.pone.0070739.

Gibson, D.J., 1996. Textbook misconceptions: The climax concept of succession. *The American Biology Teacher*, *58*(3), pp.135-140.

Glenn-Lewin, D.C., Peet, R.K. and Veblen, T.T. eds., 1992. *Plant succession: theory and prediction* (Vol. 11). Springer Science & Business Media.

Gliessman, S.R., 1992. Agroecology in the tropics: achieving a balance between land use and preservation. *Environmental Management*, *16*(6), pp.681-689.

Good, M. A., 1931, Theory of plant geography. The new phytologist vol. XXX, No. 3, Page no. 150-171.

Grotenhuis, T. P. and Bugbee, B., 1997, Super optimal CO_2 reduces seed yield but not vegetative growth in wheat. *Crop Sci.*, **37**(4): 1215-1222.

Grumbine, R. E. 1992. Ghost bears: Exploring the biodiversity crisis. Island Press. Washington, D. C. 290 p.

Grumbine, R.E., 1992. *Ghost bears: exploring the biodiversity crisis*. Island Press.

History of ecology. 2022. https://en.wikipedia.org.

https://en.wikipedia.org/wiki/Phytogeography

https://nbm.nic.in/Documents/pdf/SuccessStories/Success_Story_Bamboo_ICAR.pdf

Jensen, M. E. and Everett, R. 1993. An overview of ecosystem management principles. In Eastside forest ecosystem health assessment, Vol. 11, Ecosystem management: principles and applications. USDA Forest Service, Northern Region, Missoula, MT. pp. 7-15.

Jerry L. Hatfield, 2015, Temperature extremes: Effect on plant growth and development. *Elsevier*, **10**:4-10.

Jumba, F.R., Tibasiima, T., Byaruhanga, E., Aijuka, J., Pabst, H., Nakalanda, J.M. and Kabaseke, C., 2020. COVID 19: Lets act now: the urgent need for upscaling agroecology in Uganda (2020). *International Journal of Agricultural Sustainability*, *18*(6), pp.449-455.

Kampa, M. and Castanas, E., 2008. Human health effects of air pollution. *Environmental pollution*, *151*(2), pp.362-367.

Karthikeyan A., Deeparaj B. and Nepolean P. (2009). Reforestation in Bauxite mine spoils with *Casuarina equisetifolia* Frost. and beneficial microbes. *For. Trees Livelihoods,* **19:**153-165.

Kumar, A., Kumar, V., Bish, H. and Kumar, R., 2015. Reclamation of wasteland through different vegetative interventions. *Indian Forester*, *141*(5): 538-548.

Leff, B., Ramankutty, N. and Foley, J.A., 2004. Geographic distribution of major crops across the world. *Global biogeochemical cycles*, 18(1).

Leff, B., Ramankutty, N. and Foley, J.A., 2004. Geographic distribution of major crops across the world. *Global biogeochemical cycles*, *18*(1): 75-89.

Lindeman, R. L.,1942, The trophic-dynamic aspects of ecology. *Ecology*, 23 (4): 399-417.

Luna, R.K., 2006, *Plantation forestry in India*. International book distributors, Dehradun. Pp: 361-416.

Maiti, S.K., 2013. Ecology and ecosystem in mine-degraded land. In *Ecorestoration of the coalmine degraded lands* **(pp. 21-37).** Springer, India.

Mannering, J.V. and Griffith, D. P. 1982, Value of crop rotations under various tillage systems, Purdue University Cooperative Extension Service Agronomy Guide, AY-230.

Mason, H.L. and Stout, P.R., 1954. The role of plant physiology in plant geography. *Annual Review of Plant Physiology*, *5*(1), pp.249-270.

Merrill R. Kaufmann, Russell T. Graham, Douglas A. Boyce Jr., William H. Moir, Lee Perry., 1994, An ecological basis for ecosystem management. USDA Forest Service, General technical report RM 246, Pp:1-11.

Mitchell, C.C., Reeves, D. W. and Delaney, D. 1982, 90 years' results say yes to winter cover crops, Alabama Agricultural Experiment Station.

Mitchell, J. F. B., Manabe, S., Meleshko, V. and Tokioka, T., 1990, Equilibrium climate change and its implications for the future. Climate change*: The IPCC scientific assessment*, **131**:172.

Moolenaar, S.W., Van Der Zee, S.E. and Lexmond, T.M., 1997. Indicators of the sustainability of heavy-metal management in agro-ecosystems. *Science of the Total Environment*, *201*(2), pp.155-169.

Moss, B., 2008. Water pollution by agriculture. *Philosophical Transactions of the Royal Society B: Biological Sciences*, *363*(1491), pp.659-666.

Nair, P.K.R., 1982, Soil Productivity aspects of agroforestry. ICRAF, Nairobi.

Odum, E.P. and Barrett, G.W., 1971. *Fundamentals of ecology* **(Vol. 3, p. 5).** Philadelphia: Saunders.

Prasad, R and Rana, R., 2006. A study on maximum temperature during March 2004 and its impact on rabi crops in Himachal Pradesh. *J. of Agrometeorology*, 8(1): 91-99

Ponnuswami, R. 2019. Structure and Function of Ecosystem. Pp:1-11. https://doku.pub/ documents.

Pearson, C.J. and Ison, R.L. 1987, Agronomy of grassland systems. Cambridge University Press, Cambridge.

Pretty, J.N., 1994, Regenerating Agriculture. Earth scan Publications Ltd., London.

Rao, P.V., 2006. *Principles of Environmental science and Engineering*. PHI Learning Pvt. Ltd..

Reddy, S. R., 2011, *Principles of Agronomy*. Kalyani publishers.

Reddy, Y. T, and Reddy, G. S., 2007, *Principles of Agronomy*. Kalyani publishers.

Roth, G. W., 1996, Crop Rotations and conservation tillage, conservation tillage series #1, Penn State, College of Agricultural Sciences, CES.

Reijntjes, C.B., Haverkort and A. Waters-Bayer, 1992, Farming for the future. MacMillan Press Ltd., London.

Samra, J.S., Singh, G and Ramakrishna, Y.S. 2004. Cold wave during 2002-03 over North India and its effect on crops. The Hindu dated 10th January, 2004. p. 6.

Sarkar, D., Rakesh, S., Ganguly, S. and Rakshit, A., 2017. Management of increasing soil pollution in the ecosystem. *Advances in Research*, *12*(2), pp.1-9.

Schulze E. D., Beck, E. and Muller Hohenstein, K., *Plant Ecology*. Springer, Germany. ISBN 3-540-20833-X.

Scott Trimble, 2021, Measuring competition inn crop plants- mechanisms and outcomes. https://cid-inc.com.

Shivaramu and Shivashankar, 1994, A new approach of canopy architecture in assessing complementary of inter crops. *Indian J. Agronomy,* **39** (2): 179-189

Shukla, J. P., Amit Pandey and Pandey, K., 2009, Environmental Biology and Ecology. Narendra Publishing House. Page no. 178- 205.

Shukla, R.S. and Chandal, P.S., 1994, *Text book of Plant Ecology.* S. Chand and Company Limited, New Delli, India, pp.121-376.

Siddiqui. M. A., Parihar, S. and Bansal, T. 2018. Component II - e-Text: Agro-climatic Regions of India. http://epgp.inflibnet.ac.in.

Singh K., Pandey V.C., Singh B. and Singh R.R. (2012). Ecological restoration of degraded sodic lands through afforestation and cropping. *Ecol. Eng.,* **43:**70-80.

Singh, G., 1995, An agroforestry practice for the development of salt affected soil using *Prosopis juliflora* and *Leptochloa fusca. Agroforestry Sys.,* **27:** 61-75.

Solanki S., 1999, Rehabilitation of degraded lands through agroforestry- An overview. In compendium: Sustainable Rehabilitation of Degraded Lands through Agroforestry, NRCA, Jhansi. 1-11 pp.

Sumner, D.R., 1982, Crop rotation and plant productivity. In M. Recheigl, (ed). CRC Handbook of Agricultural Productivity, Vol. I CRC Press, Florida

Tansley, A.G., 1923. *Introduction to Plant Ecology: A Guide for Beginners in the Study of Plant Communities.* Discovery Publishing House.

Tomar O. S., 1997, Technologies of afforestation of salt-affected soils. *Int. J. Tree Crops.,* **9**: 131-158.

USDA Forest Service. 1992a. Taking an ecological approach to management. Proceedings of a workshop, Salt Lake City, UT, April 27-30, 1992.

Van Doren, D.M., Triplet, G.B. and Henry, J. E. 1975, Long term influence of tillage rotation and soil on corn Yeile, Ohio Report.

Vandermeer, J., 1989, The ecology of intercropping. Cambridge University Press, Cambridge.

Vandermeer, J., 1995, The ecological basis of alternative agriculture. Annual Review of Ecological Systems 26: 201-224.

Von Humboldt, A. and Bonpland, A., 2010. *Essay on the Geography of Plants.* University of Chicago Press.

Welch, J.R. Vincent, Auffhammer, M. Moya P.F, Dobermann, A. Dawe, D., 2010, Rice yields in tropical/subtropical Asia exhibit large but opposing sensitivities to minimum and maximum temperatures. Proc. Natl. Acad. Sci., **107**:14562-14567.

Printed in the United States
by Baker & Taylor Publisher Services